POWER TRAINS
COMPACT EQUIPMENT

FUNDAMENTALS
OF SERVICE FOS

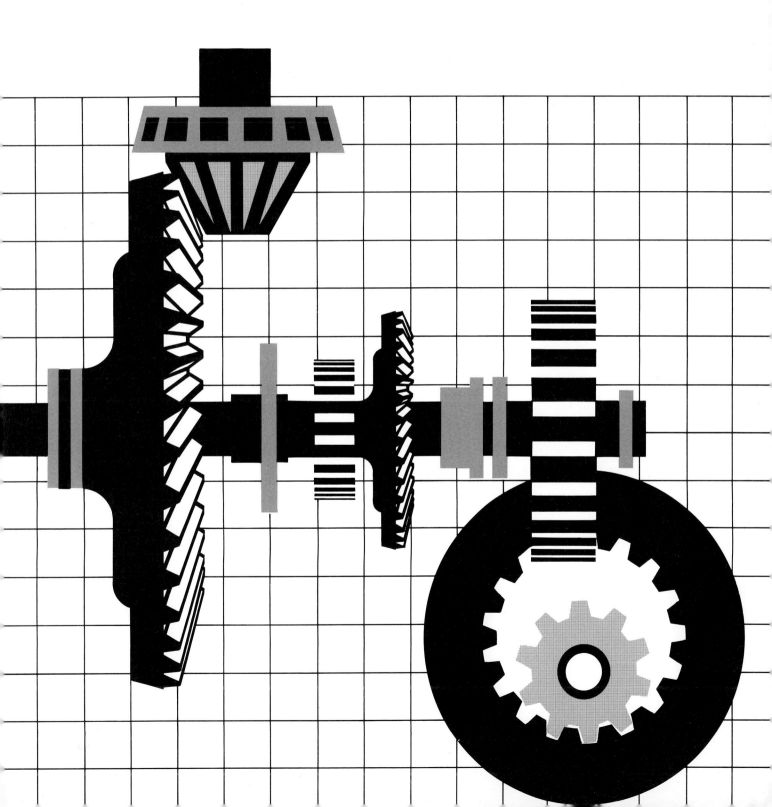

PUBLISHER

Fundamentals of Service (FOS) is a series of manuals created by Deere & Company. Each book in the series was conceived, researched, outlined, edited, and published by Deere & Company. Authors were selected to provide a basic technical manuscript which could be edited and written by staff editors.

PUBLISHER: DEERE & COMPANY SERVICE PUBLICATIONS, Dept. FOS/FMO, John Deere Road, Moline, Illinois 61265-8098; Dept. Manager: Alton E. Miller.

SERVICE PUBLICATIONS EDITORIAL STAFF
Managing Editor: Louis R. Hathaway
Editor: John E. Kuhar
Publisher: Lori J. Lees
Promotions: Cindy S. Calloway

TO THE READER:

The purpose of this manual is to help you gain greater understanding of the test and repair procedures for power trains commonly found on compact equipment.

By understanding the material covered in this manual, along with hands-on experience, you should have the basic knowledge and skills required of a beginning technician in a compact equipment service department.

COVERAGE IN THIS MANUAL

Topics covered in this manual are limited to power trains found on equipment rated up to 40 PTO horsepower (30 kW).

The power trains discussed in this manual may be found on the following equipment: chain saws, lawn mowers, riding mowers, lawn and garden tractors, compact utility tractors, skid-steer loaders, powered hole diggers, rotary tillers, and snow blowers.

To understand the test and repair procedures, it is helpful to understand how the various components of power trains work. Chapters are devoted to explanations, operating procedures, and maintenance and repair of belt, chain, and gear drives; clutches; transmissions; differentials; final drives; and power take-offs.

OTHER MANUALS IN THE SERIES

The FOS series covering compact equipment systems consists of four manuals. The three topics besides **Power Trains** are:

- **Electrical Systems**
- **Hydraulics**
- **Engines**

Each manual is backed by a Teacher's Guide, Student Workbook, and a set of 35 mm color slides.

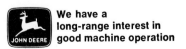

We have a long-range interest in good machine operation

FOR MORE INFORMATION

Write for a free catalog "Educational Materials: Vocational/Technical Instruction." The catalog describes each of the subjects in the compact equipment series and gives ordering instructions. The catalog also describes other subjects in the John Deere series of training materials for agricultural and industrial applications. Send your request for the catalog to:

John Deere Service Publications
Dept. FOS/FMO
John Deere Road
Moline, IL 61265-8098

ACKNOWLEDGEMENTS

The editorial staff wishes to express its gratitude to the following people who contributed significantly to the development of these books:

Frank Buckingham — Author
Donald E. Borgman — Division Sales Manager, John Deere Company, Kansas City
Bill Foster — Former Technical Services Supervisor, John Deere Horicon Works
Philip L. Lane — Division Service Manager, John Deere Company, Minneapolis
Thomas L. Shelton — Manager, Market Research, Deere & Company
Tom Brink — Service Training Supervisor, John Deere Horicon Works.

We would also like to thank the service training departments from the following John Deere units for the time many individuals devoted to the review of this manuscript: John Deere Company, Atlanta; John Deere Limited, Grimsby, Ontario, Canada; John Deere Horicon Works; John Deere Product Engineering Center, Waterloo.

Thanks also to Marson Rinkenberger of the Minnesota Curriculum Services Center in White Bear Lake, Minnesota; Thomas A. Hoerner, Professor of Agricultural Engineering and Education at Iowa State University; Robert Durland of the Extension Agricultural Engineering Dept., South Dakota State University; and David D. Schmerse, Marketing & Training at Deere & Company.

Library of Congress Catalog No. 82-71206

ISBN 0-86691-136-7

CONTENTS

720060

CHAPTER 1

POWER TRAINS TRANSFER AND CONTROL POWER

POWER TRAINS
COMPACT EQUIPMENT

FUNDAMENTALS
OF SERVICE FOS

SKILLS AND KNOWLEDGE

This chapter contains basic information that will help you gain the necessary subject knowledge required of a service technician. With application of this knowledge and hands-on practice, you should learn the following:

- **What the primary parts of power trains are.**

- **The functions of power trains.**

- **Principle means of transferring power.**

- **Components that control power.**

- **Differences between various types of clutches.**

- **Types of transmissions used.**

- **Purposes of differentials.**

- **Types and purposes of final drives.**

- **What a power take-off is.**

- **Recognizing and understanding common machine hazards.**

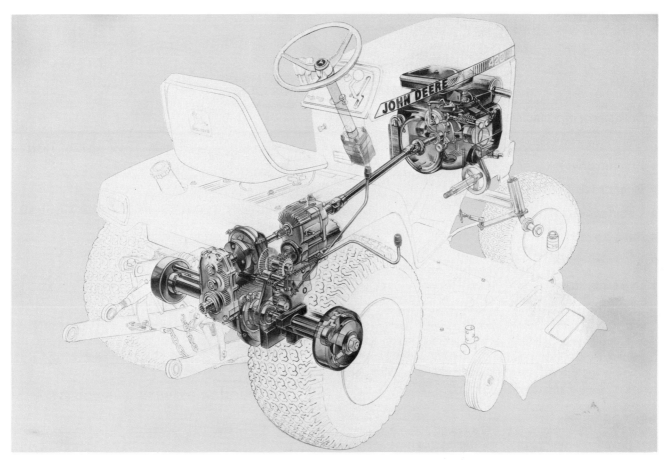

Fig. 1 — Power Trains Carry Power From The Engine To Drive Wheels Or Operating Components

INTRODUCTION

Power trains help convert engine power into useful work by:

● *Transferring power from engine to drive wheels or working components.*

● *Starting and stopping flow of power.*

● *Controlling direction and speed.*

● *Equalizing power to drive wheels for easier turning.*

Several basic parts are used to do these jobs. But they are not all included in each power train, and functions are sometimes combined in certain pieces of equipment. The basic parts and their primary functions are:

● **Shafts, belts, or hoses** to transfer power between components.

● **Clutches** to connect and disconnect power.

● **Transmissions** to select speed and direction of operation.

● **Differential** to equalize power for turning wheeled equipment such as tractors.

● **Final drives** to reduce speed and increase torque to axles.

● **Drive wheels** to carry and propel machines.

Other parts may also be used to help transfer and control power as we will see later. Also, some machines have no power train as we described it here. For instance, some power sprayers and log splitters have a pump coupled directly to the engine crankshaft, and pump operation is controlled by starting and stopping the engine. These machines are usually operated at full engine throttle, so no transmission is required to control speed.

Discussion in this book will cover equipment which permits operator control of the power flow between the engine and drive wheels or working components.

COMPONENTS THAT TRANSFER POWER

Equipment designers consider many factors in choosing parts to transfer power. Some of these factors include:

● *Amount of power to be transferred.*

● *Type of machine and normal operating conditions.*

● *Distance and direction power must be moved.*

● *Space available on or inside the machine.*

- *Need for accurate speed control or timing of components.*

- *Availability of suitable standard drive components.*

- *Machine cost.*

- *Designer or manufacturer preferences.*

Power may be transferred through:

- **Friction drives.**

- **Gear drives.**

- **Chain drives.**

- **Fluid drives.**

- **Electrical drives.**

- **Drive shafts.**

FRICTION DRIVES

Controlled friction between moving parts is widely used to transmit and control power flow in modern equipment. Common friction devices include belts and different forms of wheels and disks.

BELT DRIVES

Belt drives include two or more sheaves or pulleys and an endless round, flat, V-shaped, or specially-designed belt. V-belts are the most common type used in compact equipment. Matched sets of V-belts or two or more V-belts formed into a single unit may be used to drive heavier equipment.

Belt drives depend on tension in the belt and friction between the belt and sheaves (Fig. 2) to transfer power from one shaft to another. Some slippage between the belt and sheaves is unavoidable. But the amount of slippage will depend on the load and tension maintained in the belt. Therefore, most belt drives include a belt tightening sheave or idler. The idler may be fixed (but adjustable), or it may be spring-loaded to maintain constant tension as the belt stretches with use. A controllable belt

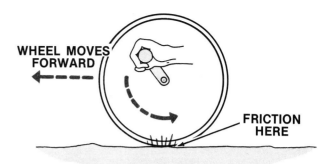

FRICTION BETWEEN WHEEL AND GROUND

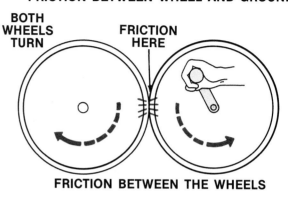

FRICTION BETWEEN THE WHEELS

Fig. 3 — Wheel Or Disk Friction

idler can be used to start and stop the power flow. Some special forms of belt drives will be discussed later.

Advantages of V-belt drives:

- Easy to design

- Inexpensive

- Absorb shock loads

- Easy to install and service

- Permit simple control of power flow

- Greater load capacity at high speeds than flat belts

Disadvantages of belt drives:

- May not last as long as some other drives

- Will not carry as heavy a load as some other drives

- Cannot be used where exact timing is required (without using special "toothed" belts)

FRICTION DISKS AND WHEELS

Friction between drive wheels and the ground propels powered machines (Fig. 3). Increasing the weight on the drive wheel, or the size of the wheel itself, and adding lugs on the wheel surface all tend to increase friction (traction) between the wheel and ground.

Fig. 2 — Belt Friction Transfers Power

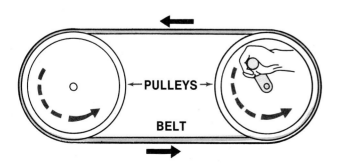

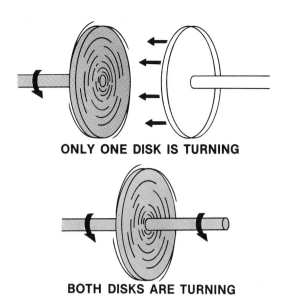

ONLY ONE DISK IS TURNING

BOTH DISKS ARE TURNING

Fig. 4 — Friction Disks Can Transmit And Control Power

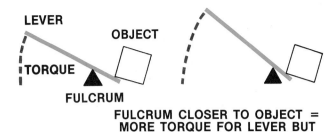

FULCRUM CLOSER TO OBJECT = MORE TORQUE FOR LEVER BUT LEVER MUST MOVE FARTHER

Fig. 5 — Torque Varies With Lever Length

Friction resulting from pressing the edges of two wheels or disks together and applying power to turn one of them will cause the other wheel or disk to rotate in the opposite direction (Fig. 3). The amount of friction developed depends largely on the surface materials and force with which the wheels are held together. Two hard, smooth surfaces such as steel would provide very little friction without excessive pressure. Therefore, most friction wheel or disk drives have coatings of rubber or other special materials with fairly high coefficients of friction (resistance to slip).

However, the contact area between the edges of friction wheels or disks is usually relatively small and this limits the amount of power which can be transferred by this type of drive. So, due to increased contact area, it is more efficient to transfer power through the faces of rotating friction disks (Fig. 4) than to use the edges of disks of the same diameter. Special materials on the disk faces increase friction and withstand heat.

GEAR DRIVES

Gears are the basic elements in the power train of most larger equipment and some smaller machines as well. They transfer power from one shaft to another and are used in transmissions, differentials, and final drives.

Basically, gears apply a twisting force called *torque* to rotating parts. Technicians should understand the principles of torque. Because of these principles high engine power can be efficiently transferred to the drive wheels of a compact tractor. Engine torque is transmitted through small,

high-speed gears in the transmission and on to the drive shaft. The small, fast-turning gears reduce torque by increasing speed so the transmission can transmit the engine power without damage. The fast-turning drive shaft then transmits torque to the large ring gear in the differential which reduces the turning speed, converting it back into torque for the drive wheels. The amount of torque developed between gears is proportional to the distance from the center of the gear to the teeth or the number of gear teeth. It is similar to the lever in Fig. 5. The lever to the right has more torque if the fulcrum is moved closer to the object. The lever must be moved further to move the object, but torque is greater.

The same principle applies to meshing gears (Fig. 6). *A small gear will drive a larger gear more slowly but with greater torque.* Providing a combination of gears to choose from gives an operator a choice of *speed* or *torque.* So:

A small gear driving a big gear (low gear) = less speed but more torque.

A big gear driving a small gear (high gear) = less torque but more speed.

Fig. 6 — Torque Varies With Relative Gear Size

SMALL GEAR DRIVING LARGER GEAR = LESS SPEED BUT MORE TORQUE

LARGE GEAR DRIVING SMALL GEAR = MORE SPEED BUT LESS TORQUE

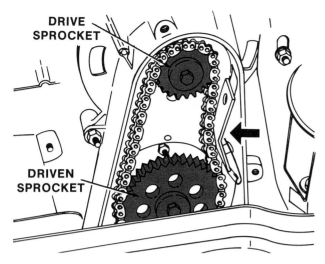

Fig. 7 — Chain Drives Provide Positive Power Flow

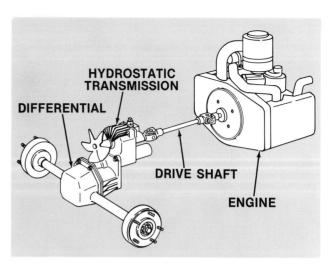

Fig. 9 — A Typical Hydrostatic Drive For A Garden Tractor

CHAIN DRIVES

Chain drives (Fig. 7) are less flexible but stronger than belt drives. They don't slip as much as belt drives, so they can be used for precision timing of moving parts.

As with belt and gear drives, the size of the driving and driven sprockets (or the number of sprocket teeth) in a chain drive determines the speed and torque of the driven shaft. For instance, the driven shaft (Fig. 7) will rotate slower than the drive shaft because the drive sprocket is smaller than the driven sprocket. The smaller sprocket must turn more than one revolution to engage enough chain links to turn the larger sprocket one complete revo-

Fig. 8 — Typical Roller Chain Design

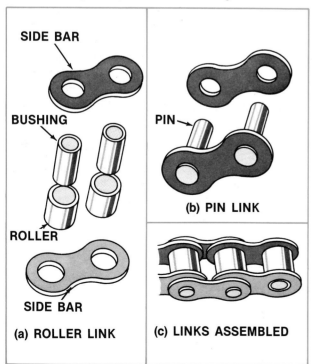

lution. Likewise, a larger drive sprocket causes the driven shaft to turn faster than the drive shaft.

There are several types of drive chain available. However, roller chain (Fig. 8) is the most common type used in the power trains of compact equipment.

FLUID DRIVES

Fluid drives are among the oldest, and also the most modern ways of transmitting power. The weight of water falling into the buckets of an old water wheel turned the wheel and powered equipment. Directing a stream of fluid, or air, against fins on a wheel causes the wheel to turn and drive other equipment.

Fluid can also transmit power from the engine to drive wheels or working components of a machine. Hydrostatic drives use high pressure oil to transmit power between a hydraulic pump and motor (Fig. 9). A hydrostatic drive can replace both the clutch and transmission in conventional drive trains and provide greater operating flexibility. It can also be used where it would be difficult or impossible to install a mechanical power train to the drive wheels.

Hydraulic motors can also be used to power mowers and other components where belts, chains, or drive shafts are impractical or undesirable.

ELECTRICAL DRIVES

Some machines are driven by electric motors which draw power from batteries installed on the vehicle. The batteries are then recharged when the machine is not in use. However, the batteries generally have a limited range of operating time and recharging batteries usually requires several hours or overnight.

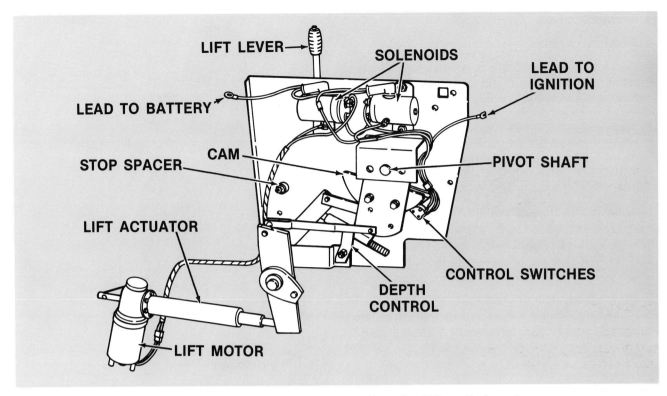

Fig. 10 — Electric Lifts Control Raising And Lowering Of Some Equipment

Electric-powered lifts (Fig. 10) are also used on some lawn and garden equipment. These systems have a small electric motor to extend and retract a screw which raises and lowers equipment such as mower decks. Such lifts replace hydraulic cylinders on equipment not equipped with a hydraulic pump.

Some vehicles also use an electric clutch in the power train to the drive wheels or for the power take-off. These clutches use electromagnets, controlled by movement of the clutch pedal or lever, to engage and disengage the clutch. Electric clutches will be discussed in more detail later.

Fig. 11 — Universal Joints Protect Drive Shafts From Misalignment

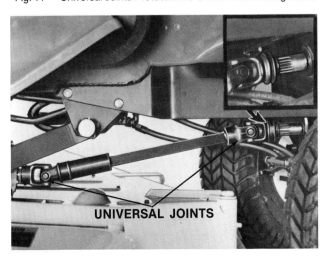

DRIVE SHAFTS

Rotary power is often transmitted between machine components by drive shafts. Depending on the load and speed of operation, the drive shaft may be tubular or solid. Tubular shafts are lighter, and in many cases will transmit as much power as a solid shaft of equal size.

It's almost impossible to maintain perfect alignment between a power source and the driven component. Therefore, most drive shafts are equipped with some form of flexible coupling or joint. This helps prevent damage to bearings. A typical flexible coupling (Fig. 11) is called a universal joint, or simply a "U-joint." Universal joints can compensate for deliberate or accidental misalignment of the drive shaft up to certain angular limits.

COMPONENTS THAT CONTROL POWER

Power flow from the engine to drive wheels or operating components of machines must be properly controlled for effective machine performance. Components that control power flow include:

- **Clutches.**
- **Transmissions.**
- **Differentials.**
- **Final drives.**

- **Hydraulic valves.**

- **Electrical controls.**

Not all these components are found in each power train. And the function of two or more of these components may be combined into a single unit in some machines.

CLUTCHES

Clutches connect and disconnect power between the source (engine) and the transmission or auxiliary mechanism such as the power take-off or mower drive.

Clutches permit the engine to run while the machine is standing still or without operating attachments such as mowers. A clutch is used in all self-propelled machines such as tractors and power mowers, except those with torque converters or hydrostatic drives.

There are many different kinds of clutches. And some designs may include features of two or more basic designs. The more common types of clutches are:

- *Belt tensioning clutches.*

- *Friction clutches.*

- *Jaw clutches.*

- *Centrifugal clutches.*

- *Electromagnetic clutches.*

Service and maintenance of these clutches will be discussed in Chapter 4.

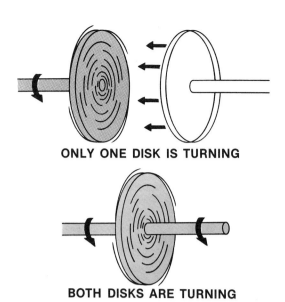

ONLY ONE DISK IS TURNING

BOTH DISKS ARE TURNING

Fig. 13 — Friction Disks Can Transmit And Control Power

Belt Tensioning Clutches

One of the simplest and most economical clutches is the belt tensioning clutch (Fig. 12). When the engine is running and the clutch is disengaged, the drive belt is loose and the drive sheave turns freely without affecting the driven sheave. However, engaging the clutch tightens the belt around both sheaves. As the belt tightens, the driven sheave starts turning and when the clutch is fully engaged (belt fully tightened), power flows to the driven sheave.

Friction Clutches

Friction clutches use mating disks, plates, bands, shoes, or cones to control power flow. The mating parts are forced together and held there by springs or centrifugal force until released by the operator. Friction between mating parts causes the driving and driven members of the clutch to rotate at the same speed.

Disk or plate clutches: As mentioned before, it's more efficient to transfer power through the faces of rotating disks than to use the disk edges. This is the operating principle behind disk or plate clutches (Fig. 13). As long as the disks do not touch each other, one can rotate without affecting the other. But, if the spinning disk is moved into contact with the other, both will spin as a single unit. In a clutch, strong springs force the disks together, and the clutch pedal acts as a lever and separates them to stop the flow of power.

Band clutches: A friction band, tightened against the outside of a rotating drum or flywheel, controls power flow in a band clutch (Fig. 14). To engage the clutch, linkage arms are drawn closer together, tightening the band. The band and flywheel then turn as a unit.

Fig. 12 — Belt Tensioning Clutches Are Simple And Effective For Small Equipment

BELT TENSIONING CLUTCH

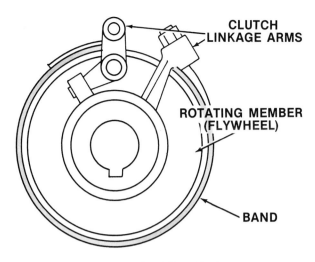

Fig. 14 — Band-Type Clutch

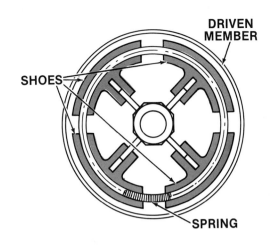

Fig. 16 — Centrifugal-Type Expanding Shoe Clutch

Because of their bulky size, band clutches are not commonly used on mobile equipment.

Expanding shoe clutches: Friction between clutch shoes and a rotating outer member transmits and provides control of power in expanding shoe clutches. Forcing the shoes against the outer members engages the clutch. Withdrawing the shoes disengages power flow.

Mechanically operated expanding shoe clutches (Fig. 15) are operated by pivoting linkage which expands the shoes against the outer member. Moving the release bearing back disengages the clutch.

Shoes in a centrifugal expanding shoe clutch (Fig. 16) are forced against the outer member by centrifugal force as the speed of the inner member increases. The faster the inner member turns, the greater the force holding shoes against the outer member. The clutch is disengaged automatically when the inner drive member slows down. These clutches are commonly used in chain saws.

Cone clutches: Forcing the mating parts of a cone clutch together transfers torque from the drive member to the driven member (Fig. 17). To increase friction, the driven member usually has a bonded lining that helps grip the drive member. A throw-out bearing on the shaft (Fig. 17) engages and disengages clutch members. Cone clutches are frequently used to control power take-offs and other auxiliary power trains.

Fig. 15 — Mechanical-Type Expanding Shoe Clutch

Fig. 17 — Cone-Type Clutch

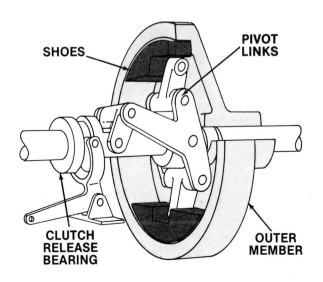

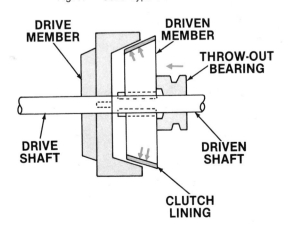

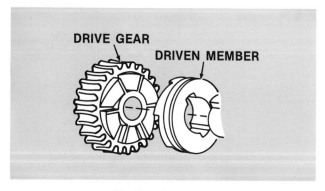

Fig. 18 — Jaw Clutch

Jaw Clutches

Engaging the members of a jaw clutch (Fig. 18) provides a positive drive with no slippage between the members. However, because clutch engagement is somewhat abrupt, jaw clutches are generally limited to relatively slow speed drives with limited power flow. The ground-drive system of walk-behind lawn mowers is one application.

Centrifugal Clutches

Centrifugal clutches rely on rotation of the drive member to develop sufficient centrifugal force to engage rollers, cams, or shoes with the driven member to transmit power. Slowing or reversing the drive member disengages the clutch and permits the driven member to freewheel or overrun the power source. This permits heavy rotating machine parts to coast to a stop without transmitting power back into the engine. A separate brake on

driven member such as lawn mower blades can be used to stop them within a safe period of time.

A roller-type centrifugal or overrunning clutch (Fig. 19) automatically engages in one direction, but freewheels in the other. As the center drive line starts turning faster, rollers tend to climb the ramps and wedge between the outer and inner members of the clutch. The members then rotate as a unit. Engagement will not occur until the drive line rotates faster than the driven member. And the drive disengages as soon as the speed of the drive member drops below that of the driven shaft. Cams, sprags, or coil springs may replace the rollers in some overrunning clutches, but clutch function is essentially the same.

Electromagnetic Clutches

Electromagnets are used to control some clutches instead of conventional sliding throw-out bearings. When the switch is actuated by moving the clutch pedal or lever, current flowing through the field coil sets up a magnetic field which draws the clutch plate against the face of the rotor assembly (Fig. 20). Power is then transmitted through the assembly the same as with a regular disk-type clutch.

When the switch is reopened, the magnetic field collapses and the face plate is freed from the rotor. This breaks the power flow immediately.

Some other electrical clutches use a mixture of magnetizable metal and powdered dry lubricant to "freeze" and "unfreeze" the clutch whenever a current is applied to or disconnected from an electromagnet. Controlled slip-

Fig. 19 — Overrunning Centrifugal Clutch

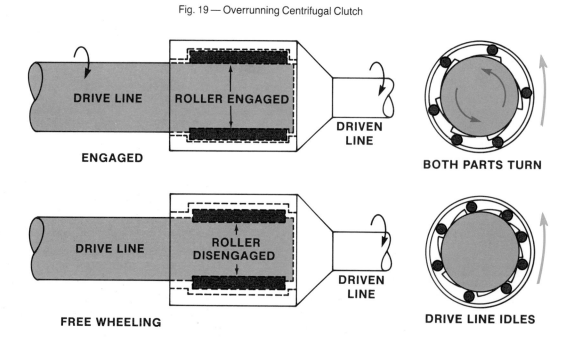

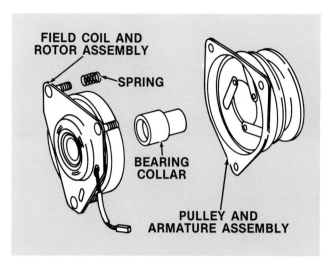

Fig. 20 — Direct-Action Electrical Control For Clutch

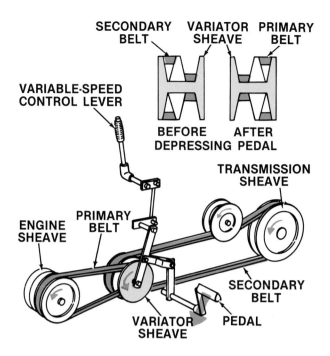

Fig. 21 — Variable-Speed Belt Drive

page or a solid drive can be provided by regulating the amount of current flowing to the electromagnet.

TRANSMISSIONS

Transmissions permit changing the speed of drive wheels or other powered components in relation to engine speed. A transmission also permits delivering more torque or power to the drive wheels at slower speeds. This is particularly important for easier starting and pulling heavy loads, or with vehicles such as automobiles and trucks.

Transmissions can also reverse power flow to drive wheels for easier maneuvering, or to other driven parts for unplugging or special operations.

Two types of transmission systems are available:

- **Mechanical transmissions**

- **Hydrostatic drives**

Mechanical Transmissions

Four types of mechanical transmissions are commonly used in various compact equipment:

- **Variable speed belt drives**

- **Dual-belt, direction or speed control**

- **Friction disk or wheel**

- **Gear**

Variable-speed belt drives: Changing the size of either sheave alters the speed of the driven shaft. Reducing the diameter of the drive sheave increases its speed. Conversely, increasing the diameter of the driven sheave reduces its speed. Also, as we discussed earlier with gears, *a smaller sheave drives a larger sheave slower but with more torque,* and *a larger sheave drives a smaller sheave faster with less torque.*

By making the sides of the sheaves adjustable, the output speed can be varied while holding engine speed constant. In a typical variable-speed drive (Fig. 21) the variator sheave has a movable center section (see inset) which can be moved in or out by a mechanical linkage. It can also be moved toward or away from the engine sheave to tighten the belt. Moving the variator sheave further from the engine sheave tightens the primary belt, forcing it closer to the center of the variator sheave thus reducing its speed (small diameter input to variator, large diameter output). At the same time, the secondary belt is forced to the outside of the variator sheave which reduces the rotating speed of the trans-

Fig. 22 — Dual Belt, Reversing Drive

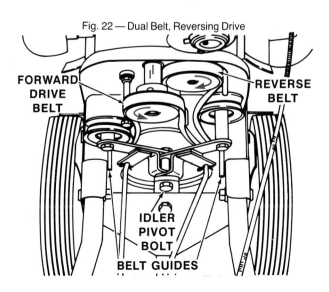

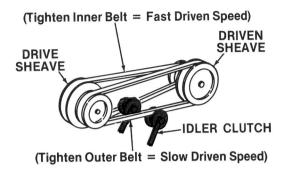

(Tighten Inner Belt = Fast Driven Speed)

DRIVE SHEAVE

DRIVEN SHEAVE

IDLER CLUTCH

(Tighten Outer Belt = Slow Driven Speed)

Fig. 23 — Dual-Belt Drive Provides Choice Of Speeds

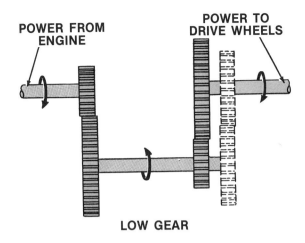

POWER FROM ENGINE

POWER TO DRIVE WHEELS

LOW GEAR

Fig. 25 — Gear-Type Transmission In Low Or First Gear

mission sheave still more. Shifting the variator sheave toward the engine results in increased drive speed to the transmission sheave.

Dual belt drives: Some walk-behind equipment such as garden tillers use two belts which permit reversing the machines under power. In a typical dual belt drive (Fig. 22), a V-belt sheave on the engine crankshaft drives the machine forward. A second sheave on an extension of the engine camshaft rotates in the opposite direction and provides power for the reverse drive. A dual sheave is mounted on the tiller tine drive shaft. When both belts are loose, the engine can run without moving the machine. Engaging the forward-drive belt idler causes the machine to move forward. When that idler is released and the reverse-drive idler is engaged, the machine moves in reverse. There is no provision for changing travel speeds.

There is a dual-belt drive arrangement which provides a choice of speeds in the same direction. Using dual sheaves of different diameters on both driving and driven sheaves (Fig. 23) and two drive belts provides two drive speeds. Engaging the idler clutch on the outer

Fig. 24 — Friction Disk Drives Can Change Speed And Direction Of Power Flow

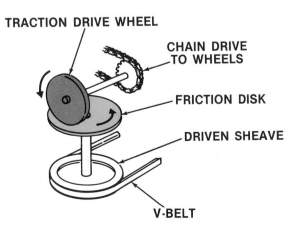

TRACTION DRIVE WHEEL

CHAIN DRIVE TO WHEELS

FRICTION DISK

DRIVEN SHEAVE

V-BELT

belt, with small drive sheave and large driven sheave, provides a slow output speed. Releasing that idler and tightening the other belt around the large drive sheave and small driven sheave provides a faster output speed.

This type of drive is found on some lawn and garden tractors in conjunction with a gear transmission to double the number of forward and reverse speeds.

Friction disk or wheel drives: Power can be transmitted through the edges of rotating disks or wheels, as discussed earlier. Turning one disk perpendicular to the other changes direction and speed of power flow.

A V-belt from the engine (Fig. 24) turns a sheave on one end of a short shaft supported by a bearing and the machine frame. Attached to the other end of the shaft is a friction disk. The disk face may be smooth or specially surfaced to increase friction between the disk and the traction drive wheel. A rubber or plastic rim on the traction wheel increases friction between the wheel and disk to provide more efficient power flow. The traction wheel can be moved laterally across the friction disk. Moving the traction wheel all the way to the outside of the friction disk provides maximum speed to the output shaft through the wheel. Shifting the traction wheel to the center of the friction disk stops the traction wheel, while moving it to the opposite edge reverses the direction of friction wheel rotation.

Gear transmissions: Several combinations of gears are provided in gear transmissions. Different combinations allow different speeds for the load and operating conditions. For low, first gear, a small gear on the input shaft drives a larger gear on the secondary shaft (Fig. 25). This drives the secondary shaft at a slower speed,

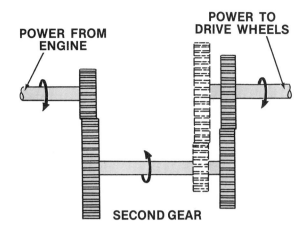

Fig. 26 — Gear Transmission In Second Gear

Fig. 28 — Liquids Have No Shape Of Their Own

but provides more torque. A small gear on the secondary shaft then drives a larger gear on the drive shaft to the vehicle's drive axle. This reduces speed still more and provides another increase in torque for starting or pulling heavy loads.

The same input gears are used for second gear (Fig. 26). But, the gears used for low gear are disconnected and the drive passes through two other gears. This time, a larger gear on the secondary shaft drives a smaller gear on the drive shaft, so output speed is faster and torque lower than in low gear. Additional gears provide more speed and torque selection.

Reverse is obtained by directing power through a reverse idler gear which causes the drive shaft to rotate in the opposite direction (Fig. 27).

All the transmission gears are mounted in a metal case which is filled with oil to lubricate and cool the gears and bearings. The gears are meshed and unmeshed by moving a shift lever.

Hydrostatic Drives

Two principles of hydraulics are involved in hydrostatic drives:

- *Liquids have no shape of their own.*

- *Liquids cannot be compressed.*

Liquids fill the shape of any container in which they are placed (Fig. 28). Therefore, hydraulic oil flows in any direction and into any open passage in the system.

To prove that liquids cannot be compressed, fill a plastic jug with liquid (Fig. 29). Then tightly cork it. If you press down on the cork, trying to compress the liquid, it won't work. But, if you press hard enough, the jug will expand and finally break.

These same principles work in hydrostatic drives as shown in Fig. 30. Two cylinders of equal size, each containing a piston, are connected by a line. The cylinders

Fig. 27 — Gear Transmission In Reverse Gear

Fig. 29 — Liquids Cannot Be Compressed

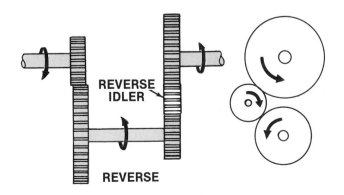

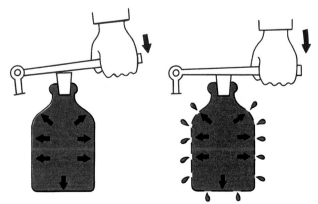

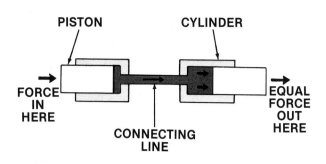

Fig. 30 — Connected Cylinders Transmit Forces Through A Fluid

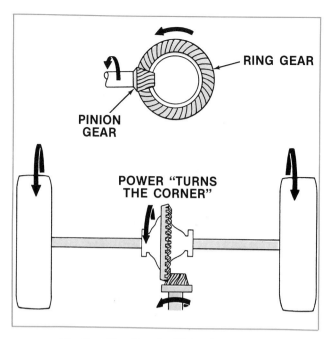

Fig. 32 — Ring Gear And Pinion For Drive Axle

and lines are filled with oil. Applying a force to the left piston moves that piston against the oil. The oil will not compress, so it acts as a solid and forces the right piston to move.

In a hydrostatic drive, several pistons in a pump transmit power to a second set of pistons in a motor (Fig. 31). Varying the rate, pressure, and direction of oil flow between the pump and motor controls the speed, power output, and direction of operation of the hydrostatic drive. Hydrostatic drive controls permit infinite variations in oil pressure and rate of flow between full reverse and full forward speeds. This gives an endless selection of speed and torque for smooth, convenient operation.

Hydrostatic drives can be used where it would be difficult or impossible to install a mechanical drive train.

DIFFERENTIALS

Machines with a drive shaft to carry power from the transmission to the drive axle use a ring gear and pinion to transmit power around the corner to the axle (Fig. 32).

Fig. 31 — Typical Hydrostatic Drive Pump And Motor

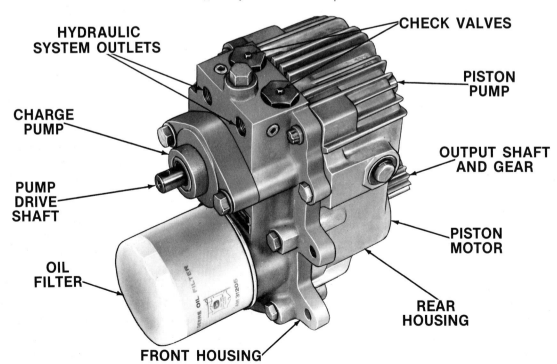

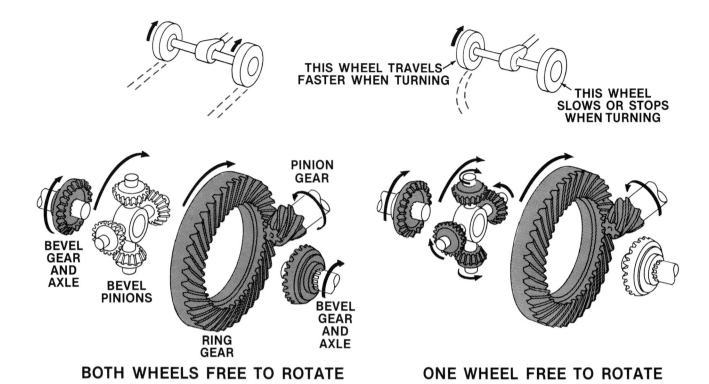

Fig. 33 — Operation Of A Differential

However, when any machine turns a corner, the wheel on the outside of the turn must travel further than the inner wheel. This also means the outer wheel rotates faster than the inner wheel during the turn. The differential (Fig. 33) permits each wheel to travel at a different speed and still propel its portion of the load.

Power from the pinion gear rotates the ring gear which is meshed with the four bevel pinions. When the machine moves straight ahead, each axle receives equal power and both wheels rotate at the same speed (left, Fig. 33). When the machine turns a sharp corner, the inner wheel (on the right, Fig. 33) slows or stops. The bevel pinions are then forced to "walk around" the right-hand bevel gear. This turns the left-hand bevel gear faster because more turning force from the ring gear is transmitted to it through the bevel pinions.

If the right-hand wheel stops completely, the left-hand bevel gear (and axle) make two complete revolutions for each turn of the ring gear; one as the ring gear turns and another as the bevel pinions "walk around" the other bevel gear.

The differential also permits the wheel with the least resistance to turn faster. Thus if one drive wheel hits a slick spot and spins, the other one will turn slower by the same amount.

Some machines have a differential lock or limited-slip differential to lock the wheels together so power is kept on the wheel with traction so it can pull the machine out of slick spots. When the traction equalizes on both wheels, the lock releases.

FINAL DRIVES

The system which transfers power from the differential to the drive wheels is called the final drive. There are four main types of final drive systems:

- **Straight axle**
- **Pinion**
- **Planetary**
- **Chain**

In addition, on some machines, a hydrostatic drive motor is mounted directly in the drive wheel hub, eliminating a mechanical power train between the engine and wheels.

Straight-axle final drives carry power directly from the differential to the drive wheels (Fig. 34). However, pinion, planetary, and chain final drives usually provide an additional speed reduction and increase in torque before power reaches the drive wheels.

A speed reduction at this point permits lighter-weight power train components which operate at higher speeds. How? Power transferred is the product of speed and torque in the drive. If either of these is

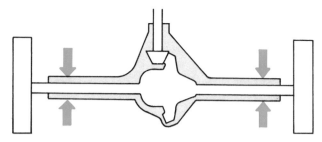

RIGID AXLE SHAFT

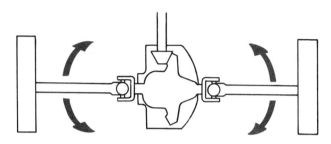

FLEXIBLE AXLE SHAFT

Fig. 34 — Two Types Of Straight-Axle Final Drives

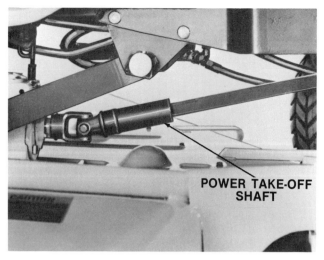

POWER TAKE-OFF SHAFT

Fig. 35 — Power Take-Off Drive

changed and the other held constant, power increases or decreases in direct proportion. Consequently, large amounts of power can be transmitted through relatively small (lightweight and lower cost) components rotating at higher speed. When drive speed in the final drive is reduced, torque is increased. There are minor power losses in each bearing, gear set, and chain in the power train, but these are usually more than offset by the advantages of higher speed drive trains.

POWER TAKE-OFFS

The power take-off delivers power from the power train to auxiliary equipment. This equipment may be mounted on or drawn by the vehicle supplying the power.

A power take-off is usually a flexible drive shaft with universal joints which couples a splined shaft on the vehicle to the implement drive train (Fig. 35). However, for some machines, power may be transferred through belts. And in others, a hydraulic motor is used instead of a mechanical drive shaft.

Power take-off (PTO) drives must be properly shielded to prevent accidental injury of the operator or bystanders in case they contact the rotating shaft. On some equipment, a switch under the seat automatically

Fig. 36 — Three Types Of Hydraulic Valves

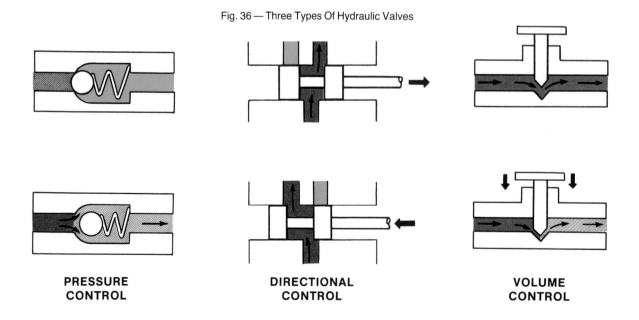

PRESSURE CONTROL

DIRECTIONAL CONTROL

VOLUME CONTROL

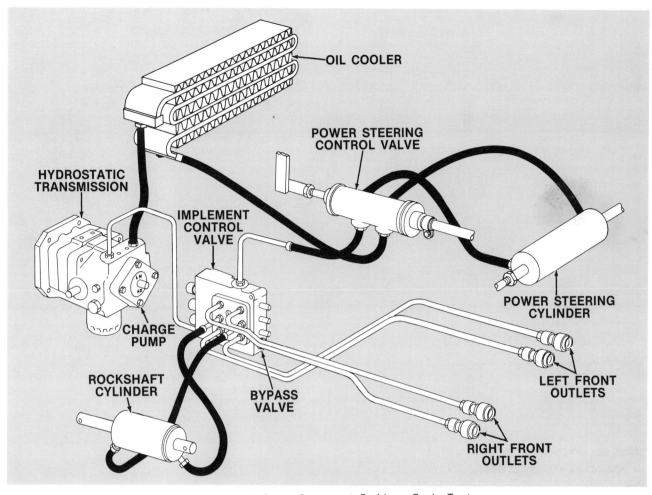

Fig. 37 — Hydraulic System Components On A Large Garden Tractor

shuts off the engine if the operator leaves the seat without disengaging the PTO clutch.

HYDRAULIC CONTROLS

The flow of hydraulic power is usually controlled by hydraulic valves. Depending on the design and complexity of the system, hydraulic valves (Fig. 36) may be used to control volume, start, stop, or redirect the flow of oil.

Hydraulics play an important role in the operation of machines (Fig. 37). However, except for vehicles with hydrostatic drive or hydraulically actuated clutches, the hydraulic system is not considered an integral part of the vehicle power train. Therefore, only hydrostatic drives and hydraulic clutches are mentioned here. For more information on hydraulic systems refer to publications listed in the Suggested Readings in the Appendix as well as the operator's manual and technical service manual for the machine concerned.

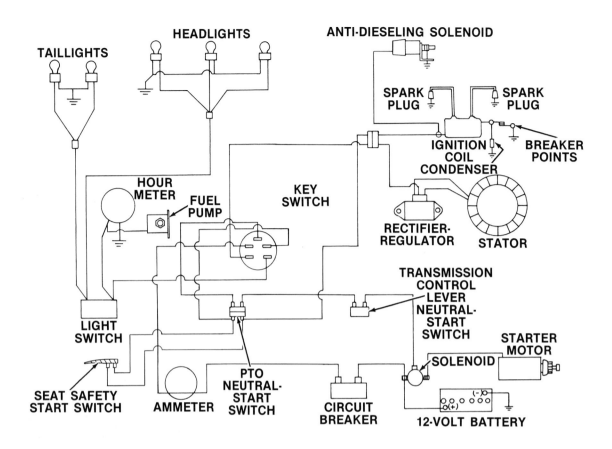

Fig. 38 — Typical Vehicle Electrical System

ELECTRICAL CONTROLS

With the exception of instruments and controls (Fig. 38), and battery-powered electric vehicles, electromagnetic clutches are the only important electrical component in compact equipment power trains, and these will be discussed later.

For more information on vehicle electrical systems refer to publications listed in the Suggested Readings in the Appendix as well as the operator's manual and the technical manual for the machine.

COMMON MACHINE HAZARDS

This section deals with machine hazards. It will help you recognize hazards and understand why they are hazardous. You will know what to avoid. The information in this section will help you develop a safe attitude.

Here are some of the most common machine hazards you should recognize:

• **Pinch points**

• **Wrap points**

• **Thrown objects**

• **Stored energy**

• **Slips and falls**

PINCH POINTS

Pinch points are where two parts move together and at least one of them moves in a circle (Fig. 39). Pinch points are also referred to as mesh points, run-on points, and entry points. There are pinch points in power transmission devices such as belt drives, chain drives, and gear drives (Fig. 39).

You can be caught and drawn into the pinch points by loose clothing. Contact with pinch points may be made if you brush against unshielded rotating parts, or slip and fall against them. A person can get tangled in a pinch point if he or she deliberately takes chances and reaches near rotating parts. Machines operate too fast for a person to withdraw from a pinch point once they are caught.

AVOIDING PINCH POINTS

Manufacturers build shields for pinch points. Always replace shields if you must remove them to repair or adjust a machine. Remember that a por-

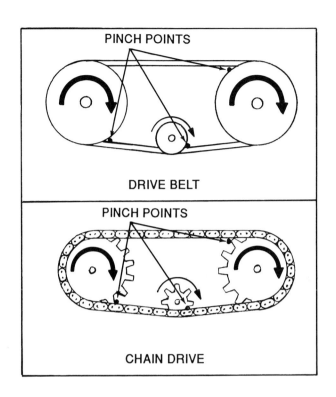

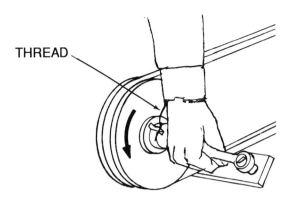

THREAD

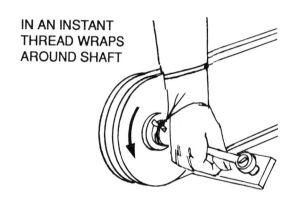

IN AN INSTANT
THREAD WRAPS
AROUND SHAFT

Fig. 39 — Pinch Points On Rotating Parts Can Catch Clothing, Hands, Arms, And Feet

SLEEVE IS
IMMEDIATELY
PULLED
AND
BEGINS
TO
WRAP

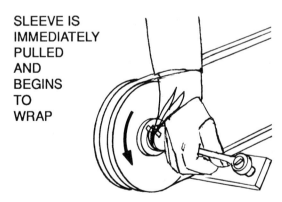

tion of the money you paid for the machine went for safety research and design and for the actual hardware involved in shielding. Get your moneys worth and protect yourself.

For pinch points that cannot be shielded, the best protection available is operator awareness. Know the location of the pinch points on your machinery. Avoid them when the machine is operating. And above all, never attempt to service or unclog a machine until you have disengaged all power, shut off the engine, removed the key, and all parts have stopped moving.

WRAP POINTS

Any exposed component that rotates is a potential wrap point. Rotating shafts are usually involved in wrap-point accidents. Often, the wrapping begins with just a thread or frayed piece of cloth, catching

Fig. 40 — Wrapping May Begin With Just A Thread. In An Instant The Victim Is Entangled With Little Chance To Escape Injury

on the rotating part (Fig. 40). More fibers wrap around the shaft. There's no escape.

The shaft continues to rotate pulling you into the machine in a split second. *The more you pull the tighter the wrap.* If the clothing would tear away, a person might escape serious injury, but work clothes are usually too rugged to tear away safely. Long hair can also catch and wrap, causing serious, permanent injury.

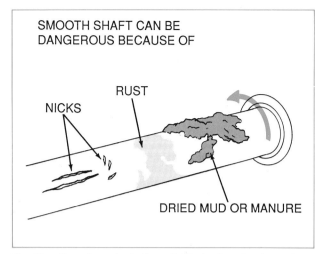

Fig. 41 — Even Seemingly Smooth Shafts Can Catch And Wrap Clothing

Smooth shafts often appear harmless, but they too can wrap and wind clothing. Rust, nicks, and dried mud or manure make them rough enough to catch clothing (Fig. 41). Even shafts that rotate slowly must be regarded as potential wrapping points.

Fig. 42 — Shafts That Extend Much Beyond Bearings Or Sprockets Can Be Dangerous

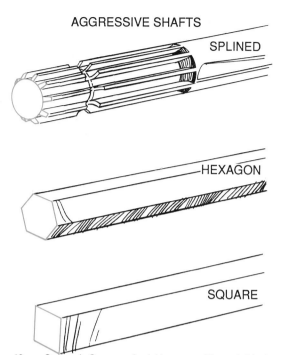

Fig. 43 — Splined, Square, And Hexagon-Shaped Shafts Are More Likely To Entangle Than Round Shafts

Ends of shafts which protrude beyond bearings can wrap up clothing (Fig. 42). Splined, square and hexagon-shaped shafts are more likely to wrap than smooth round shafts (Fig. 43).

Couplers, universal joints, keys, keyways, pins, and other fastening devices on rotating components will wrap clothing (Fig. 44).

THROWN OBJECT HAZARDS

Some machines can throw objects great distances with tremendous force (Fig. 45). Recognize machines with a throwing hazard so you can avoid injury.

AVOIDING THROWN OBJECT HAZARDS

1. *Recognize what machines throw objects.*

2. *Keep the machines properly shielded* to reduce the possibility of thrown objects. Some operations may require removing or adjusting shields.

Make sure you replace the shielding before you begin the next job.

Follow the manufacturer's instructions for shielding to reduce the hazard of thrown objects.

3. *Know how far and in what direction objects may be thrown,* even with shielding in place (Fig. 46).

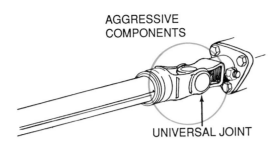

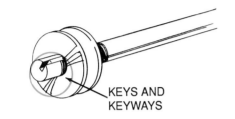

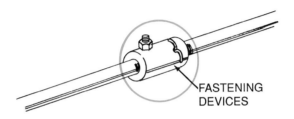

Fig. 44 — Some Components Are Even More Aggressive Than Some Shafts

Fig. 46 — Know How Far And In What Direction A Machine May Throw Objects. Use Shields As Operation Permits

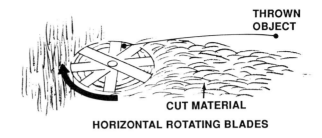

Fig. 45 — Stones And Sticks Are Thrown Farther And Harder Than Grass

4. *Stay a safe distance away from the likely path of thrown objects when you approach a machine.*

5. *When operating a machine which may throw objects, make sure the machine will not discharge near people or animals. Build a shield if necessary.*

STORED ENERGY

Stored energy can work for you, or it can be carelessly released and cause injury.

On compact equipment, energy is stored so it can be released at the right time, in the right way for you. Here are some components that store energy. You should recognize them as you use and service compact equipment:

- **Springs**
- **Hydraulic Systems**
- **Compressed Air**
- **Electricity**
- **Raised Loads**
- **Loaded Mechanisms**

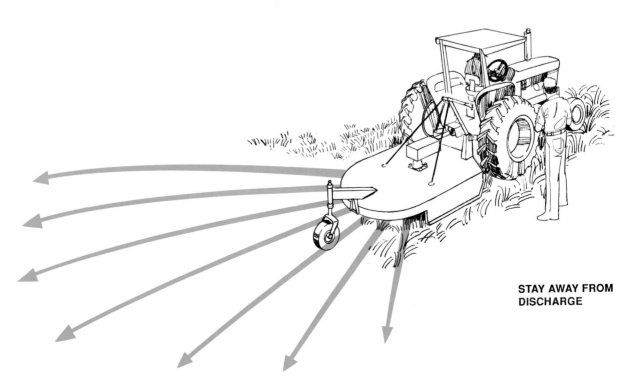

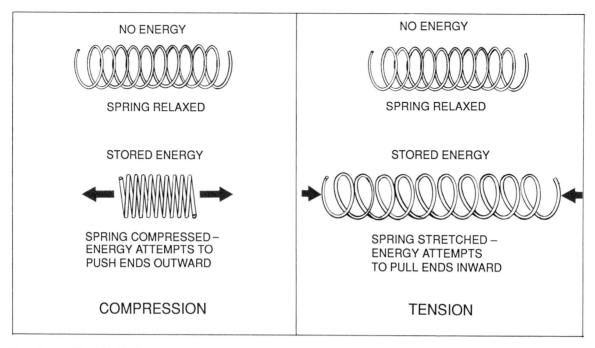

NO ENERGY

SPRING RELAXED

STORED ENERGY

SPRING COMPRESSED—
ENERGY ATTEMPTS TO
PUSH ENDS OUTWARD

COMPRESSION

NO ENERGY

SPRING RELAXED

STORED ENERGY

SPRING STRETCHED —
ENERGY ATTEMPTS
TO PULL ENDS INWARD

TENSION

Fig. 47 — Energy Stored In Springs

SPRINGS

Springs are energy storing devices. They are used to help lift implements, to keep belts tight, and to absorb shock. Springs store energy in tension or in compression (Fig. 47).

When you remove any device connected to a spring, be sure you know what can happen. Know what direction the spring will move, and what direction it will move other components when it is disconnected (Fig. 48). A compressed spring can propel an object outward with tremendous force.

Make sure you and others will not be in the path of any part that will move when the spring moves. Plan exactly how far and where each part will move. Use proper tools to assist you in removing or replacing spring-loaded devices. Even small springs can store a lot of energy.

Fig. 48—Be Sure You Know What Can Happen Before Disconnecting Any Part Attached To A Spring

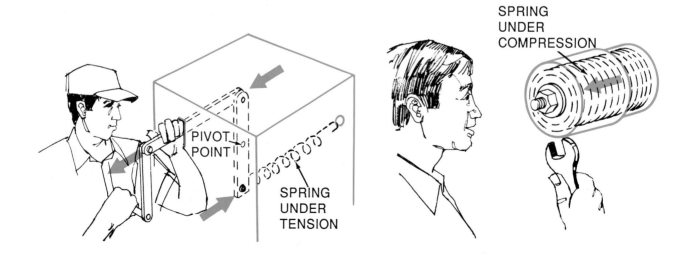

PIVOT POINT

SPRING UNDER TENSION

SPRING UNDER COMPRESSION

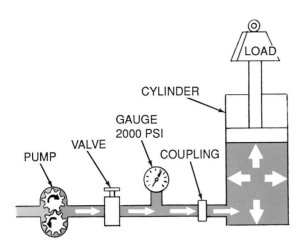

Fig. 49 — Hydraulic Fluid Under Pressure Attempts To Escape Or Move To A Point Of Lower Pressure

HYDRAULIC SYSTEMS

Hydraulic systems on compact equipment also store energy. Hydraulic systems must confine fluid under high pressure often higher than 2,000 pounds per square inch (13,790 kPa).

A lot of energy may be stored in a hydraulic system, and because there is often no visible motion, operators do not recognize it as a potential hazard. Carelessly servicing, adjusting, or replacing parts can result in serious injury. Fluid under pressure attempts to escape (Fig. 49). In doing that it can do helpful work, or it can be harmful.

Servicing and Adjusting Systems Under Pressure

Adjusting and removing components when hydraulic fluid is under pressure can be hazardous (Fig. 50). Imagine attempting to remove a faucet from your kitchen sink without relieving the water pressure. You'd get a face full of water! It is much more dangerous with hydraulic systems. Instead of just getting wet from water at 40 psi (276 kPa), you would be seriously injured by oil under 2,000 psi (13,790 kPa), or more. You could be injured by the hot, high pressure spray of fluid and by the part you are removing when it is thrown at you (Fig. 50).

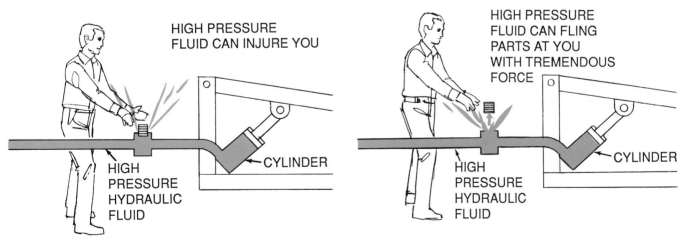

Fig. 50 — Always Relieve Hydraulic Pressure Before Adjusting Hydraulic Fittings. You Could Be Injured By A Hot, High Pressure Spray Of Hydraulic Fluid Or By A Part Flung At You

AVOID HIGH-PRESSURE FLUIDS

Escaping fluid under pressure can penetrate the skin causing serious injury. Avoid the hazard by relieving pressure before disconnecting hydraulic or other lines. Tighten all connections before applying pressure. Search for leaks with a piece of cardboard. Protect hands and body from high-pressure fluids.

If an accident occurs, see a doctor immediately. Any fluid injected into the skin must be surgically removed within a few hours or gangrene may result.

Fig. 51 — Avoid High-Pressure Fluids

To avoid this hazard, always relieve the pressure in a hydraulic system before loosening, tightening, removing, or adjusting fittings and components. Keep all hydraulic fittings tight to prevent leaks. But, do not tighten fittings without relieving the pressure. Also, if you over-tighten a coupling, it may crack and release a high-pressure stream of fluid. You will be injured by the fluid and the implement, which will drop to the ground.

Before attempting any service:

1. *Shut off the engine which powers the hydraulic pump.*

2. *Lower implement to the ground.*

3. *Move the hydraulic control lever back and forth* several times to relieve pressure.

4. *Follow instructions in operator's manual.* Specific procedures for servicing hydraulic systems are very important for your safety.

Trapped Oil

Hydraulic oil can be trapped in the hydraulic system even when the engine and hydraulic pump are stopped (Fig. 52). Trapped oil can be under tremendous pressure, up to 2,000 psi (13,790 kPa), or more. You can be seriously injured by escaping fluid and moving machine parts if you loosen a fitting.

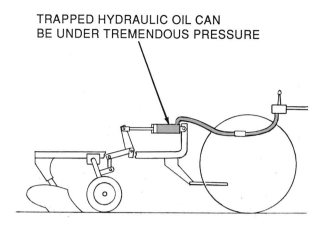

TRAPPED HYDRAULIC OIL CAN BE UNDER TREMENDOUS PRESSURE

Fig. 52 — Even Though The Engine And Hydraulic Pump Are Stopped, Hydraulic Oil Can Be Trapped Under Tremendous Pressure — A Potential Hazard

Another hazard with trapped oil (Fig. 53) is heat. Heat from the sun can expand the hydraulic oil and increase pressure. The pressure can blow seals and move parts of an implement or machine.

Fig. 53 — Trapped Oil Can Move An Implement And Blow Seals

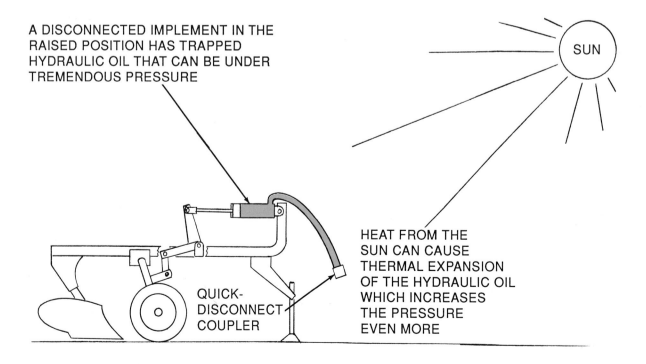

A DISCONNECTED IMPLEMENT IN THE RAISED POSITION HAS TRAPPED HYDRAULIC OIL THAT CAN BE UNDER TREMENDOUS PRESSURE

QUICK-DISCONNECT COUPLER

SUN

HEAT FROM THE SUN CAN CAUSE THERMAL EXPANSION OF THE HYDRAULIC OIL WHICH INCREASES THE PRESSURE EVEN MORE

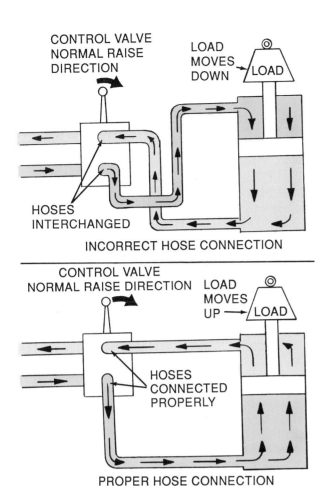

Fig. 54 — Don't Interchange Hydraulic Lines

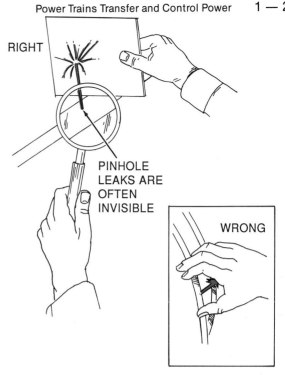

Fig. 55 — The Jet Stream Or Mist From A Pinhole Leak In A Hydraulic System Can Penetrate Your Skin — Don't Touch It!

Incorrect Coupling

Crossing hydraulic lines creates hazards. When lines are coupled to the proper part, you get the results you expect. But if lines are crossed, the implement may raise when you expect it to drop (Fig. 54). Serious injury could result. Make sure hydraulic lines are coupled exactly as specified in the machine technical manual. Color code lines with paint or tape. After you attach hydraulic lines try the controls cautiously to see if you get the proper result.

Fig. 56 — Don't Connect A High-Pressure Hydraulic Pump To Low-Pressure Systems

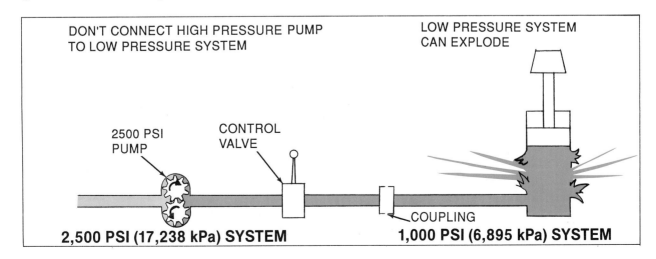

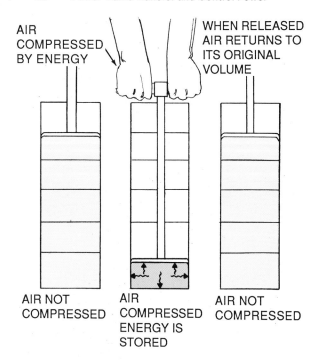

AIR COMPRESSED BY ENERGY

WHEN RELEASED AIR RETURNS TO ITS ORIGINAL VOLUME

AIR NOT COMPRESSED

AIR COMPRESSED ENERGY IS STORED

AIR NOT COMPRESSED

Fig. 57 — When Air Is Compressed, Energy Is Stored. The Air Attempts To Return To Its Original Volume

Do not replace hydraulic pipes with rubber hose. Pipes are usually designed to carry heavier pressure than rubber hose. Hose is usually not as durable as pipes.

Another hazard is coupling a high-pressure pump to a low-pressure system. Attaching a hose from a 2,500 psi (17,238 kPa) hydraulic system to an implement equipped with hoses, cylinders, and fittings designed for 1,000 psi (6,895 kPa) is inviting trouble (Fig. 56). The low pressure system could explode. Never improvise or adapt couplings or fittings on a low-pressure system to attach to a high-pressure pumping system. Always follow the manufacturer's recommendations.

Pinhole Leaks

If liquid, under high pressure, escapes through an extremely small opening it comes out as a fine stream (Fig. 55). The stream is called a *pinhole leak.* Pinhole leaks in hydraulic systems are hard to see and, they can be very dangerous. High-pressure streams from pinhole leaks penetrate human flesh. Hydraulic systems on some compact equipment may have pressures of 2,000 psi (13,790 kPa) or higher. That's higher than the pressure in hydraulic syringes used to give injections. Injury from pinhole hydraulic leaks comes from the fluid cutting through flesh, and from body reaction to chemicals in the fluid.

You may see only the symptoms of pinhole leaks from high pressure systems. There may appear to be only a dripping of fluid, when actually it may be an accumulation of fluid from a high-pressure jet stream so fine it is invisible. Don't touch a wet hose or part with bare or even gloved hands to locate the leak. Pass a piece of cardboard or wood over the suspected area instead (Fig. 55). Wear safety glasses. Then relieve the pressure and replace the defective part.

Fig. 58 — Compressed Air Is Stored Energy. Always Stand To One Side When Inflating Tires And Never Over-Inflate Them

Diesel fuel injectors are designed to force fuel into engine cylinders under high pressure. Don't touch the jet stream from a diesel injector nozzle. It is just as dangerous as a pinhole leak.

⚠ **CAUTION: Escaping fluid under pressure can penetrate the skin causing serious injury. Avoid the hazard by relieving pressure before disconnecting hydraulic or other lines. Tighten all connections before applying pressure. Search for leaks with a piece of cardboard. Protect hands and body from high pressure fluids. If an accident occurs, see a doctor immediately. Any fluid injected into the skin must be surgically removed within a few hours or gangrene may result.**

COMPRESSED AIR

Compressed air is dangerous. When air is compressed, its volume is reduced (Fig. 57). Energy is stored.

Wear safety glasses when using compressed air to clean. Never use more than 30 psi (207 kPa).

Probably the most frequent use of compressed air for compact equipment is for tire inflation. An inflated tire can be dangerous, especially the large ones. When air is compressed into a tire, it attempts to get out.

When that large quantity of compressed air gets an opening to return to its original volume, perhaps through a failure of the tire or its seal, there's a tremendous amount of stored energy released. A 10.00-20 12PR truck tire inflated to 75 psi (517 kPa) has 46,510 foot-pounds (63,068 N.m) of energy, enough energy to raise a 3000 pound (1,361 kg) car 15 feet (4.6 m). A 24.00-49 tire inflated to 75 psi (517 kPa) develops 354,260 foot-pounds (480,377 N.m) of energy, which could lift a 134-pound (61 kg) person one-half mile (0.8 km) into the air!

You should recognize that compressed air is stored energy. It can be hazardous if not properly controlled and respected. Keep pressure at proper levels. Stand to one side when inflating tires (Fig. 58). Follow manufacturer's recommendations at all times. For more detailed information on tire care, maintenance, and safety refer to "Fundamentals of Service — Tires and Tracks." John Deere FOS manual series.

CAUTION: Every tire and rim or wheel must be handled in a special way. Always use the tire and rim and wheel manufacturer's procedures when you demount and mount a tire.

Information also can be obtained from the following associations:

Rubber Manufacturer's Association
1400 K Street, N.W.
Washington, D.C. 20005

National Wheel and Rim Association
4836 Victor Street
Jacksonville, FL 32207

Further safety information can be obtained from:

U. S. Department of Transportation
National Highway Traffic Safety Administration
400 Seventh St., S.W.
Washington, D.C. 20590

ELECTRICAL ENERGY

One of the most common forms of stored energy is electricity stored in 12 volt batteries. When properly used it makes your work easier. If handled carelessly it can cause serious injury. If you recognize the potential hazards of electricity, you'll be able to avoid serious accidents.

Fires

Every self-propelled machine should have a 5 pound (2.3 kg) all-purpose, ABC dry chemical fire extinguisher on board.

Electrical systems can cause fires if not properly maintained. The energy stored in the battery may be tapped to start the engine. But if a bare wire touches a metal part and becomes hot or sparks, it can start a fire in dust, chaff, and leaves. Most machinery fires do not result in personal injury, but every fire is a potential source of injury. Inspect electrical systems. Make sure wires are properly insulated and clean dust, chaff, leaves and oil off wires.

Short-Circuit Starting

If insulation on electrical wires is cracked or worn, a short circuit occurs. Electricity could flow to the cranking motor and start the engine when no one is around. If the positive and negative terminal of a cranking motor are accidentally contacted by another metal object such as a wrench, the current will flow between the two terminals, and accidentally start the engine.

DANGER

Do not short across starter with a screwdriver to start a tractor. You bypass the neutral start switch by doing so, and the tractor can lunge and crush you.

In summary, the energy in an electrical system is waiting to do something. If it does what was planned for it, at the right time, there's no problem. If it does what was planned for it at the wrong time or does the wrong thing injury and

Fig. 59 — Use The Steps, Ladders And Handholds To Safely Get On And Off Compact Equipment — Don't Jump

property damage result. Inspect electrical systems on all machines. Replace worn wiring, contacts and switches.

SLIPS AND FALLS

Slipping and falling can put a person out of commission for hours, days, or years. Slips and falls can be prevented by recognizing the potential hazards and avoiding them.

ON-OFF ACCIDENTS

Compact equipment is equipped with steps, handholds, and platforms that help you get on and off (Fig. 59). Steps are made so you can have three support points resting on the machine at all times. Two feet and one hand, or two hands and one foot. Use the hand holds and steps, and keep them in good repair. Don't jump off machines.

Recognize what can interfere with steps, handholds, and platforms. Your good judgment can prevent accidents.

SLIPPERY FOOT SURFACES ON MACHINES

Mud, snow, ice, manure, and grease may build up on steps, platforms, and other surfaces. When it does, you can slip and fall (Fig. 60). Or, you could slip and bump a control, causing the machine to lurch into action, injuring yourself or someone else. Also, you could fall into a moving part. Such falls can be fatal.

Take time to clean foot surfaces for your own safety and for others who will use them. Also, wear boots with non-skid soles.

Fig. 60 — Mud, Snow, Manure, And Grease On Foot Surfaces Of Machines Could Cause A Serious Fall

CLUTTERED STEPS AND OPERATOR PLATFORMS

Chains and tools on operator platforms are accidents waiting to happen. You need to be able to move about without having to watch where you place your feet. Slipping on mud or snow may be somewhat excusable, but tripping on something you placed on a step isn't. Machine surfaces intended for your feet should be kept clear for your feet (Fig. 61).

SLIPPERY GROUND SURFACES

When there's snow, ice, or mud on the ground you can't do much to change it. But you can learn to recognize that those slippery surfaces can lead to accidents and try to avoid them.

The danger is usually slipping or tripping and falling against or into a machine that is running. When you must work under these conditions, the best practice is to slow down, step deliberately, and be on the lookout for slippery surfaces or objects that may cause you to lose your footing. Also, wear boots with non-skid soles.

CHAPTER 1 REVIEW

1. Name four functions of power trains.

2. Name the basic parts of a power train.

3. Name at least five principle means of transferring power.

4. (Choose one) If a belt drive sheave is larger than the driven sheave, speed of the *driven* shaft will be (*faster, slower*) than the speed of the drive shaft.

Fig. 61 — Keep Machine Platforms Clear

5. How is power transferred through a belt drive?

6. (Fill in the blank) _____ are the basic elements in many power trains.

7. (Choose one) A small gear driving a larger gear will deliver (*more, less*) torque to the driven shaft.

8. Why is a clutch needed in a power train?

9. What is the purpose of a transmission in a power train?

10. What two principles of hydraulics are involved in hydrostatic drives?

11. What is torque?

12. How can most common machine hazards be avoided?

13. (Fill in the blank.) You must _____ common machine hazards in order to avoid them.

14. True or False? "Smooth, rotating shafts are harmless."

15. Why can't a person whose clothing is caught tear away from wrap points?

16. What are four parts that store energy?

17. (Fill in the blanks.) You should _____ the pressure of a hydraulic system before attempting to service it.

18. True or False? "The hydraulic system of an implement disconnected from the tractor does not contain high-pressure hydraulic oil."

19. Where should you stand when inflating a tire?

20. Name two hazards of electrical systems on compact equipment.

21. How can slips and falls be prevented?

CHAPTER 2

FEATURES OF BELT, CHAIN, AND GEAR DRIVES

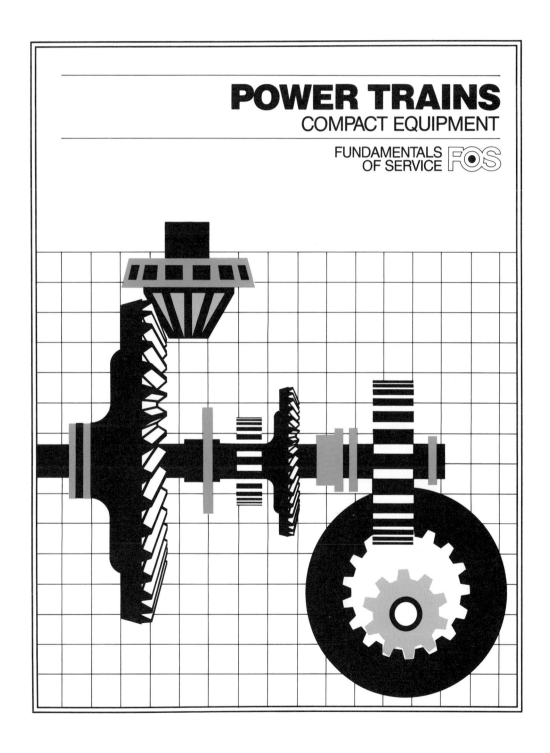

POWER TRAINS
COMPACT EQUIPMENT

FUNDAMENTALS
OF SERVICE F●S

SKILLS AND KNOWLEDGE

This chapter contains basic information that will help you gain the necessary subject knowledge required of a service technician. With application of this knowledge and hands-on practice, you should learn the following:

- **General features of belt drives.**

- **Care of belts.**

- **Construction of V-belts.**

- **Alignment of belt drives.**

- **Adjusting belt tension.**

- **Troubleshooting belt drives.**

- **Different types of chain.**

- **Operation of chain drives.**

- **Chain alignment and proper tensioning.**

- **Repairing chain drives.**

- **Operation of gear drives.**

- **Types of gears and their purposes.**

- **Operation of planetary gears.**

- **Gear backlash and endplay and how to adjust.**

- **Types of gear wear.**

- **Types of bearings and their uses.**

- **Power train lubrication.**

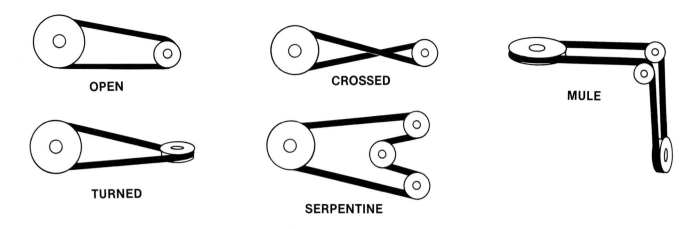

OPEN   CROSSED   MULE

TURNED   SERPENTINE

Fig. 1 — Forms Of Belt Drives

INTRODUCTION

Belts, chains, and gears are used in compact equipment power trains to transmit and control the flow of power from the engine to drive wheels or working components such as mower blades.

Understanding how these drives function and how to care for them is important. It is also important to understand the function and care of the bearings and lubrication used in these drives.

BELT DRIVES

The ability of belts to transmit power depends on several variables:

● *Tension holding the belt to the sheaves.*

● *Friction between belt and sheaves.*

● *Arc of contact area (or "wrap") between belt and sheaves.*

● *Speed of the belt. (Higher speeds reduce tension and contact.)*

Although belts are normally used to transmit power between two parallel shafts, they can be used to drive more shafts and can be arranged in a variety of configurations (Fig. 1).

CARE OF BELTS

Belts of different types — flat, round, special-shaped, and "V" — all require similar care and maintenance. However, V-belts are the most common and the following applies particularly to them.

1. Store belts in a cool, dry place. Coil long belts in natural loops when storing them on pegs. Avoid sharp bends in belts and do not hang belts over nails or other small diameter supports.

2. Relieve tension on belts when not in operation.

3. Never force a belt onto a sheave (Fig. 2). Forcing damages cords which may in turn damage the sheaves. Most drives have movable sheaves or idlers to ease belt installation. If not, remove the sheaves to remove or install a belt.

4. Do not expose belts to lubricants. Grease and oil cause a belt to slip and soften and it weakens the belt structure. Also, do not use belt dressing to reduce V-belt slippage. Instead, increase belt tension or replace the belt to eliminate excessive slippage.

5. Avoid overloading belts. Excessive loads cause abnormal slippage and rapid wear.

Fig. 2 — Never Force A Belt Onto A Sheave

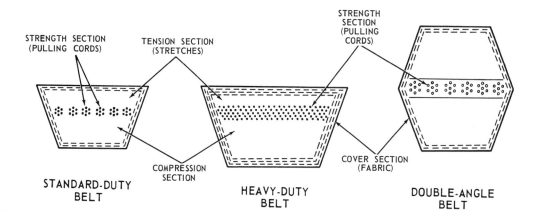

Fig. 3 — Construction Of Common V-Belts

V-BELT CONSTRUCTION

V-Belts are made in several types (Fig. 3), but each one is made up of four sections.

The top section of the belt is known as the *tension section* (Fig. 3). It is rubber and *stretches* as the belt forms around the sheave.

The bottom section is called the *compression section* because it *compresses* when wedged into and shaped around the sheave.

The center section of the belt is known as the *strength section.* It neither compresses nor stretches, but the cords located in this area give the belt its *tensile strength.* Without these cords, a V-belt would be pulled in two.

The complete belt is covered by a *cover section* of tough fabric and rubber which protects the inner parts. (Some belts — called "raw edge" have no fabric cover on the sides.)

Notice in Fig. 3 that the *heavy-duty* belt uses a different material for the cords and may use a heavier fabric cover than the *standard-duty* type.

Double-angle V-belts are used for drives where both the top and the bottom of the belt must contact sheaves as in a serpentine drive. In these belts, the strength cords are in the center of the belt and either side and absorb compression and tension forces.

Some V-belts also have corrugated undersides for greater flexibility.

Most V-belts are endless, but open-end types are produced for use where a closed belt cannot be installed. The ends have metal fasteners joined by a pin, or a link and two pins.

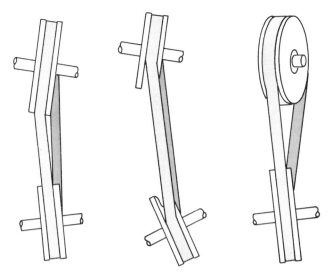

Fig. 4 — Sheave Misalignment Can Damage V-Belts

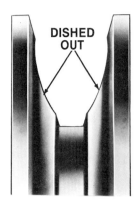

Fig. 6 — Sheave Grooves "Dished Out" By Wear

BELT ALIGNMENT

Improper sheave alignment is less obvious with V-belts than with flat or round belts. Misalignment (Fig. 4) can still cause uneven belt wear, belt roll-over, cord stretch, or separation because the load is shifted to one side of the belt.

To check sheave alignment, place a straightedge next to the sheaves or stretch a cord across the sheaves (Fig. 5). The cord should touch the sheaves at each of the four arrows shown, and the shafts must be parallel. If the sheaves do not line up, loosen one or both sheaves and move them on the shaft until they align. Then retighten the sheaves on their shafts. Next, rotate each sheave a quarter turn at a time and look to see whether the contact of either sheave with the straightedge or cord is disturbed. If a sheave doesn't touch the cord or straightedge at one of the points, the shaft is bent or the sheave is wobbling because it is bent.

A bent shaft or damaged sheave will increase belt wear and could damage shaft bearings. If the problem is severe, vibration could lead to damage or failure of other machine parts. Replace or repair bent or damaged sheaves and shafts immediately.

BELT TENSION

Flat belts must have more tension than V-belts, but only enough tension to avoid slippage. Too little belt tension causes slippage or slip-and-grab operation which can break the belt. If the belt doesn't break, slippage can cause excessive wear, burned spots, and overheating.

However, too much tension causes the belt to heat up and stretch and possibly damage sheaves, shafts, and bearings. Refer to the machine operator's manual for proper tension on each belt.

Unless the drive is equipped with a spring-loaded idler, tighten belts when slippage occurs or the drive fails to work properly from lack of tension.

Check the tension of a new belt after the first hour of operation. Watch for excessive slippage and wear during the first 24 hours of use when the initial seating and stretch occurs.

Fig. 5 — Checking Alignment Of Belt Sheaves

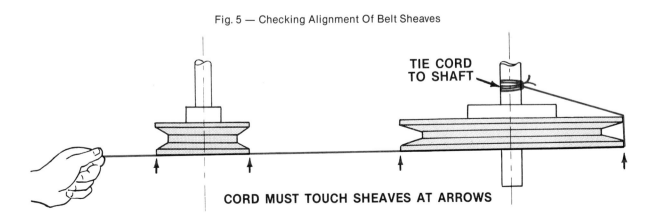

CORD MUST TOUCH SHEAVES AT ARROWS

Remember, V-belts should ride on the two sides of the sheaves; not in the bottom of the groove. If the bottom of the groove is shiny, it is a sign the belt or sheave is badly worn or the belt is too small. If the groove is shiny, replace the belt.

If the groove of a V-belt sheave is "dished out" by wear (Fig. 6), repair or replace the sheave. The grooves wear rapidly when exposed to abrasive dust or chemicals in the air or corrosion and mois-ture accumulated while the machine is idle for long periods.

Drive belts may fail due to cracking, rupturing, tear-ing, burning, gouging, excessive wearing, internal cord failure, cuts and peeling, and fraying and chewing. For detailed information on each of these types of failures, refer to "Fundamentals of Service — Identification of Parts Failures."

TROUBLESHOOTING BELT DRIVES

In case of a belt drive malfunction, look for trouble signs and correct problems **before** replacing belts and re-suming operation.

Trouble	Possible Causes
Belts turn over in sheaves	1. Misaligned sheaves and shafts
	2. Worn sheave grooves
	3. Misalignment of idler sheave
	4. Excessive belt vibration due to improper tension
	5. Belt cord damage from improper installation
Belt squeaks or screeches	1. Too high starting load, causing belt to slip because of improper tension
	2. Overload, causing belt slippage because of improper tension
Belt chirps	1. Movement of belt on flat idler or sheave (not harmful)
Belt breaks prematurely	1. Foreign material in sheaves
	2. Shock or extreme overload
	3. Belt damaged during installation
Belt stretched beyond take-up	1. Drive operated with too much tension
	2. Wrong belt or poor storage in damp area
Belt has short life	1. Worn sheaves
	2. Oil or grease on belt
	3. High temperatures
	4. Belt cover wear caused by guard or belt guide interference
	5. Excessive belt slippage (inadequate tension)
	6. Poor storage

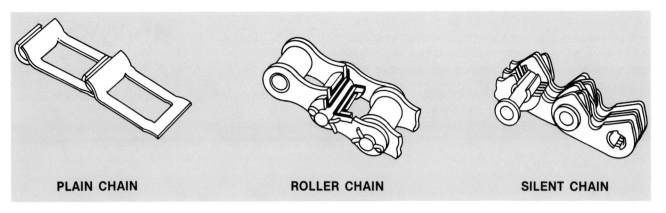

Fig. 7 — Three Types of Chain

| PLAIN CHAIN | ROLLER CHAIN | SILENT CHAIN |

CHAIN DRIVES

Chain drives generally are stronger than belt drives. Chains also eliminate slippage. But, they are less flexible and more noisy than belt drives.

Three types of chains are commonly used. Each is different to accommodate different loads and operating conditions. The three types (Fig. 7) are:

● **Plain or detachable link** — Generally used for slow speed operation. Links can be readily detached.

● **Roller chain** — Adaptable to a wide variety of load and operating conditions. Roller chain is made up of alternate roller links and pin links (Fig. 8). Bushings are free to turn. Double pitch or extended pitch roller chains have longer side plates and one roller for every other sprocket tooth. They are lighter and less expensive than regular roller chains.

● **Silent chains** — Quieter with less vibration than roller chains. They can be operated at higher speeds. Sprockets resemble gears and links are small in proportion to the chain's strength.

CHAIN DRIVE OPERATION

Some of the principles of chain drive operation (Fig. 9) are as follows:

● Chain drives transmit power from one rotating shaft to another.

● Chain drives consist of two or more sprockets and an endless loop of chain.

● Chain links mesh with sprocket teeth and maintain a positive speed ratio between the driving and driven sprockets.

● Sprockets turning on the same side of a chain revolve in the same direction. Those on the other side of the chain revolve in the opposite direction (see Fig. 9).

● Roller chain sprockets should have 10 or more teeth to avoid excessive wear.

Fig. 8 — Typical Roller Chain Construction

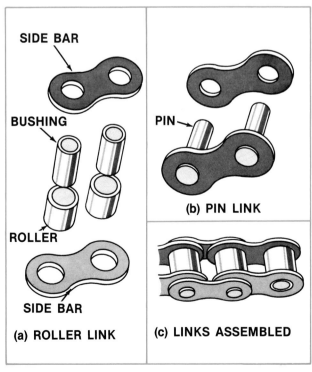

SIDE BAR

BUSHING

PIN

ROLLER

SIDE BAR

(a) ROLLER LINK

(b) PIN LINK

(c) LINKS ASSEMBLED

Fig. 9 — Principle Parts Of A Chain Drive

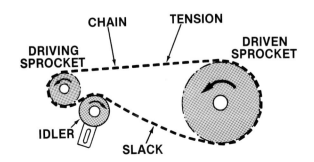

CHAIN   TENSION

DRIVING
SPROCKET

DRIVEN
SPROCKET

IDLER

SLACK

• A chain with an even number of pitches (spaces between links) or sprockets with an uneven number of teeth prevents a single pitch from contacting the same tooth each revolution. This avoids uneven wear and vibration.

• Small sprockets cause sharper bending and more chain wear.

• Shorter links bend less and should be used on smaller diameter sprockets.

• Chains may be installed in single or multiple strands, depending on the load.

• Chains must be properly tensioned by moving one of the main sprockets or the idler sprocket, or by using a spring-loaded tightener.

• Chains should have the slack side that comes *off* the drive sprocket on the bottom if possible.

• Idlers should be on the slack side of the chain.

ADVANTAGES AND DISADVANTAGES

Compared to other drives, chains offer both advantages and disadvantages. Some *advantages* of chain drives are as follows:

• Chain drives are efficient and do not slip.

• Chain drives are fairly flexible and compact.

• Chain drives are reasonably inexpensive.

• Chains keep a fixed speed ratio without slipping.

• Chains withstand heat and dirt better than some other types of drives.

• Chain are less affected by weather.

• Chain drives can carry heavier loads than belts.

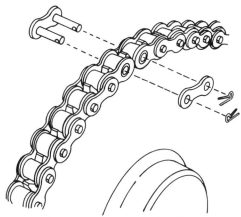

Fig. 11 — Installing Roller Chain

Some disadvantages of chain drives include the following:

• Chain drives are fairly noisy compared to belt drives.

• Most chains require frequent lubrication.

• Most chains will tolerate very little misalignment.

• Chains cannot be used if the drive must slip without adding a slip clutch or overload protector.

CHAIN ALIGNMENT

Chain sprocket shafts must be parallel and the sprockets must be aligned. To check alignment, place a straightedge against the sides of both sprockets (Fig. 10) and move sprockets into alignment if necessary. When aligned, tighten sprockets on the shaft. Be sure idlers are aligned too, and adjust if necessary.

To install a roller chain:

1. Wrap the chain around the sprocket.

2. Bring the free ends together on one sprocket.

3. Install a connector link (Fig. 11).

4. After installing fasteners, tap the chain pin links back so that fasteners are snug against the outside of the connecting link plates. This helps three ways:

• *Proper clearance is maintained between link plates so oil can move between the plates.*

• *Joints will flex freely for smooth chain action.*

• *Chain fasteners will last longer.*

CHAIN TENSION

Proper tension of drive chains avoids excessive chain and sprocket wear and possible damage to shafts and bearings. Running chains too loose lets links ride up on sprocket teeth and, in severe cases, even jump teeth. If the chain is too tight, unnecessary stress is placed on shafts and bearings as well as chain links.

Fig. 10 — Aligning Chain Sprockets

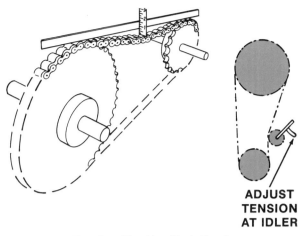

ADJUST
TENSION
AT IDLER

Fig. 12 — Checking Chain Tension

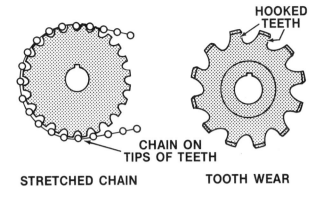

HOOKED
TEETH

CHAIN ON
TIPS OF TEETH

STRETCHED CHAIN   TOOTH WEAR

Fig. 14 — Wear Problems With Chain Drives

Horizontal and inclined drive chains should sag about $1/2$ inch per foot (19mm per M) between shaft centers (with one side of chain taut). Adjust both spans almost taut on vertical drives and those subject to shock loading or reversal. Depending on machine design, tension may be adjusted by moving one of the shafts (be sure to keep shafts parallel) or by resetting a movable idler sprocket or shoe.

To measure chain tension, pull one side of the chain taut so that all slack accumulates in the opposite span. Then place a straightedge over the slack span (Fig. 12) and measure the sag from the straightedge to the chain.

CHAIN REPAIRS

To separate roller chain:

1. Remove the fasteners and side plate from a pin link.

2. Carefully, drive out the pins with alternate strokes. Avoid damaging adjacent links.

3. If the pins have been riveted, carefully grind off the heads before trying to remove the link.

A chain-breaking tool (Fig. 13) permits quick, efficient link removal.

Prolong sprocket life by reversing sprockets if the design permits. Check alignment carefully after making the change.

Do not insert new links in a chain elongated by wear. The new links will have a shorter pitch causing a shock load each time they engage the sprockets. This will soon destroy the chain. Also, never install new chain on badly worn sprockets (Fig. 14) or worn chain on new sprockets. This can cause more chain damage in a few hours than would happen in years of normal use.

Fig. 13 — Chain Detaching Tools

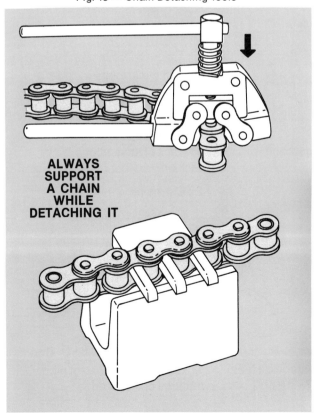

ALWAYS
SUPPORT
A CHAIN
WHILE
DETACHING IT

Fig. 15 — Cleaning Chain In Diesel Fuel

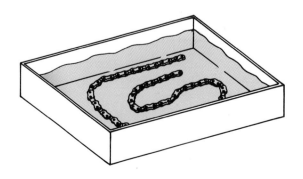

Cleaning a chain periodically will extend chain life and improve its performance. To clean a chain:

1. Remove chain from sprockets.

2. Wash chain in diesel fuel or similar solvent (Fig. 15).

DO NOT USE GASOLINE BECAUSE OF POSSIBLE FIRE OR EXPLOSION!

3. Drain fuel from chain and soak it in oil (several hours or overnight).

4. Hang chain to drain excess lubricant before replacing on sprockets.

Chain drives may fail due to excessive wear, fracture and corrosion. For detailed information on each of these types of failures, refer to "Fundamentals of Service — Identification of Parts Failures."

TROUBLESHOOTING CHAIN DRIVES

Trouble	Possible Cause
Excessive noise	1. Misalignment of sprockets
	2. Improper tension
	3. Lack of lubrication
	4. Loose bearings
	5. Chain or sprocket worn out
	6. Chain pitch too large
Wear on side bars, link plates, and sides of sprocket teeth	1. Misalignment of sprockets or idler
Chain climbing sprockets	1. Wrong chain or poor quality chain
	2. Worn chain or sprockets
	3. Lack of chain wrap on sprocket
	4. Too much chain slack
	5. Material buildup in sprocket tooth pockets
Broken pins, bushings, or rollers	1. Chain runs too fast
	2. Heavy slack or sudden loads
	3. Material buildup in sprocket tooth pockets
	4. Lack of lubrication
	5. Chain or sprocket corrosion
	6. Wrong chain or worn sprockets
Chain fasteners fail	1. Vibration
	2. Obstructions striking fasteners
	3. Fasteners improperly installed
Drive runs too hot	1. Chain running too fast
	2. Lack of lubrication
	3. Chain or shafts rubbing against obstruction
	4. Shaft bearings worn out
Chain clings to sprocket	1. Wrong chain or sprockets
	2. Heavy or tacky lubricant
	3. Excessive chain slack
	4. Material buildup in sprocket tooth pockets
Chain whips	1. Too much slack in chain
	2. High, pulsating loads
	3. Stiff chain joints (inadequate lubrication)
	4. Uneven wear on chain
Chain gets stiff	1. Lack of lubrication, resulting in wear
	2. Corrosion
	3. Excessive overloads
	4. Material buildup in chain joints
	5. Peening of side plate edges when recoupling chain
	6. Misalignment
Broken sprocket teeth	1. Obstruction or foreign material entering drive

2. Excessive shock loads

3. Chain climbing sprocket teeth (inadequate tension)

GEAR DRIVES

Gears are the most common way to transmit power for heavy loads, and they provide smooth power transfer in minimum space.

GEAR DRIVE OPERATION

There is no slippage between properly meshing gears in a power train. Therefore, the speed of the driven shaft depends on the speed of the driving shaft and the number of teeth on each gear.

For instance, if the driving gear (Fig. 16) has 12 teeth, and the driven gear 24 teeth, one complete revolution of the drive gear results in the meshing of 12 teeth on each gear. The driven gear, however, has made only one-half a revolution. This means the driven shaft turns at half the speed of the drive shaft. On the other hand, a 24-tooth drive gear making one revolution would turn a 12-tooth driven gear two complete turns and the driven shaft would rotate at twice the speed of the drive shaft.

TYPES OF GEARS

Gear shafts may be in line, parallel, or at an angle to each other. But, *meshing gears must have teeth of the same size and design,* and at least one pair of teeth must be engaged at all times. Some tooth designs permit simultaneous contact between more than one pair of teeth.

Gears are normally classified by:

● *Type of teeth.*

● *The surface where teeth are cut (edge, beveled edge, side of gear).*

Two major types of gear teeth are *straight cut* and *helical cut* (Fig. 17). The most common types of gears are shown in Fig. 18.

Straight spur gears have teeth cut straight across the perimeter, parallel to the axis of rotation. One to two pairs of mating teeth are engaged at all times. But, these gears tend to be noisy and are used mainly for slow speeds to avoid excessive vibration.

Uses — Straight spur gears are used in sliding gear transmissions because they are easily shifted by sliding on the shaft from one mating gear to another.

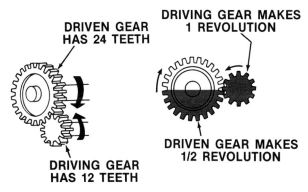

DRIVEN GEAR HAS 24 TEETH

DRIVING GEAR MAKES 1 REVOLUTION

DRIVING GEAR HAS 12 TEETH

DRIVEN GEAR MAKES 1/2 REVOLUTION

SMALL GEAR ALWAYS TURNS FASTER

Fig. 16 — Output Speed Depends On Relative Gear Size

Helical spur gears have teeth cut diagonally across the perimeter of the gear. Engagement starts at the leading tips of the teeth and rolls down the teeth to the trailing edge. The angular contact causes side thrusts which must be absorbed by bearings. However, helical spur gears are quieter and stronger than straight cut gears because there is more tooth area in contact for the same size gear.

Uses — Helical spur gears are widely used in transmissions because they are quieter at high speeds.

Fig. 17 — Two Major Types Of Gear Teeth

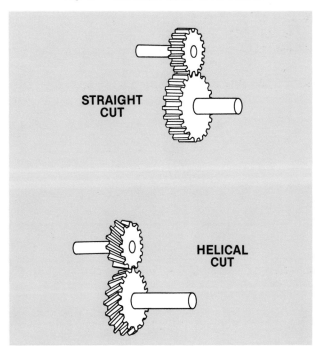

STRAIGHT CUT

HELICAL CUT

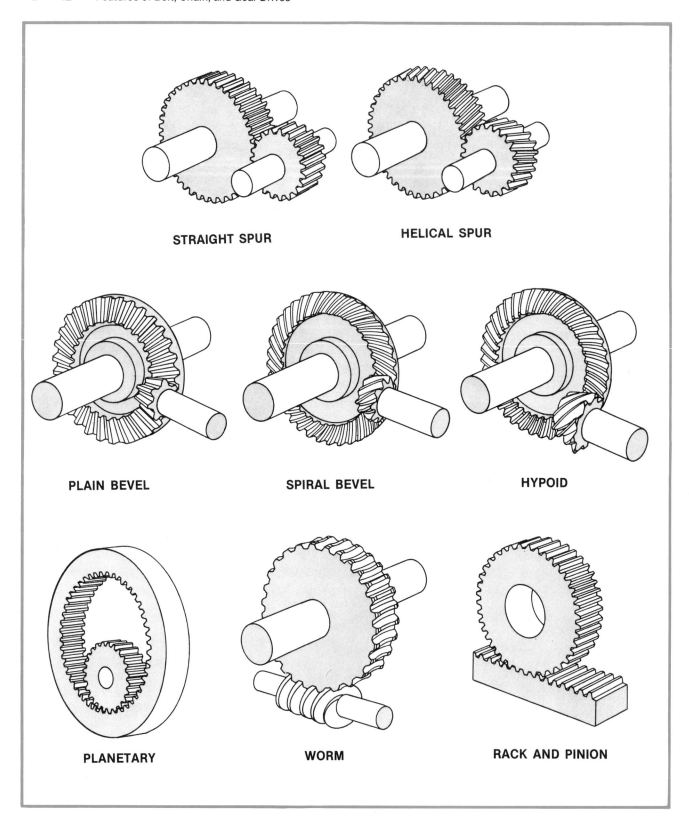

STRAIGHT SPUR   HELICAL SPUR

PLAIN BEVEL   SPIRAL BEVEL   HYPOID

PLANETARY   WORM   RACK AND PINION

Fig. 18 — Types of Gears

Plain bevel gears permit power flow to "turn a corner." Gear teeth are cut straight on a line with the shaft but at some angle between perpendicular and parallel to the shaft. The larger, driven gear is commonly called a "ring gear" and the smaller, driving gear is usually known as a "pinion gear."

Uses — Plain bevel gears are commonly used in slow-speed applications not subject to high impact forces.

Spiral bevel gears were developed for higher speed and strength in changing the angle of power flow. Teeth are cut obliquely on the angular faces of the gears. The angle is determined by the angle between the two shafts.

Uses — Spiral bevel gears are widely used in drive axle and gear-and-pinion sets of farm and industrial equipment. They not only change angle of power flow, but can also reduce speed and increase torque.

Hypoid gears resemble spiral bevel gears but the smaller pinion drive gear is located below the center of the ring gear. Teeth and general construction are otherwise the same.

Uses — Hypoid gears are mainly used in automobile differentials to lower the transmission drive shaft, thus permitting lower body styles.

Worm gears are actually a screw (inclined plane), capable of high speed reductions in a compact space. Teeth on the mating gear are curved at the tips to provide greater contact area. Power is supplied to the worm gear, which drives the mating gear. Worm gears usually provide right-angle power flows. Worm gear type drives are very inefficient.

Uses — Worm gears are often used with a high speed source to provide slow-speed, high-torque output. Small power hand tools often use a high-speed motor with a worm gear drive. Many steering mechanisms use a worm gear on the steering shaft (from the steering wheel) and a partial sector gear connected to steering linkage.

Rack and pinion gears convert straightline motion into rotary motion and vice versa.

Uses — Rack and pinion gears are commonly used for slow-speed applications such as machine tools.

Planetary gears are actually gear sets in which an outer ring gear with internal teeth meshes with smaller planet gears. The planet gears also mate with a central sun gear. Many changes in speed and torque are possible by braking or driving different parts in the set.

Uses — Planetary gears are widely used in final drives of heavy machinery and in transmissions. Each set is capable of more than one speed change. The load is spread over several gears, reducing stress and wear on any one gear.

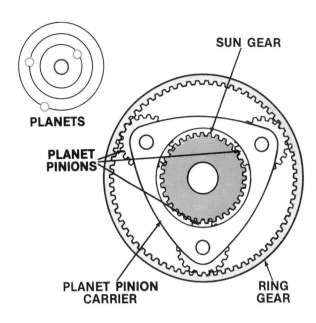

Fig. 19 — Simple Planetary Gear System

HOW PLANETARY GEARS WORK

Planetary gears have many applications and their operation should be well understood. Actually, planetary gear systems are similar to our solar system (Fig. 19). The *planet pinion gears* each turn on their own axis while rotating around the sun gear, much like the earth and other planets rotate around the sun. In turn, the pinion gears mesh with the inside of the ring gear. Notice that the sun gear, planet pinions, and ring gear are constantly in mesh.

The planet pinions are mounted on shafts in the carrier, and can rotate on their own axes to "walk around" the sun gear and/or the ring gear. Applying power to turn the sun gear, or the planet carrier, causes the entire system to rotate unless a restraining force such as a brake is applied to hold one of the other two members of the system stationary. When one member of the planetary system is powered, and a second member is restrained from turning, the remaining part becomes a power output source as illustrated.

When the sun gear is driven (Fig. 20) and a brake is applied to the ring gear, planet pinions "walk around" the ring gear, forcing the planet pinion carrier to rotate in the same direction as the sun gear, but at a slower speed.

Driving the planet pinion carrier and braking the ring gear (Fig. 21) causes the pinions to "walk around" the ring gear, forcing the sun gear to rotate in the same direction but at a higher speed.

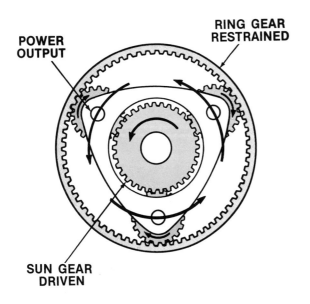

Fig. 20 — Power Flow In A Planetary When Sun Gear Is Driven

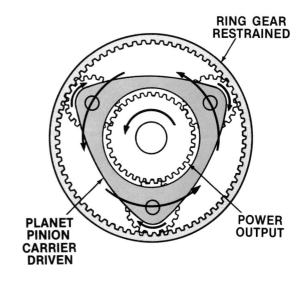

Fig. 21 — Power Flow When Planet Pinion Carrier Is Driven

Fig. 22 — Double Plant Pinion Set Gives Reverse Speed Too

Adding a second set of planet pinions (Fig. 22) to the simple planetary system, so that the two sets of pinions are in mesh, permits reversing the drive. Applying power to the planet pinion carrier (Fig. 21) and braking the ring gear causes planet pinions in mesh with the ring gear to rotate on their axes, which drives the inner planet pinions and in turn forces the sun gear to turn in the reverse direction from the planet pinion carrier. Such a system will provide low, high, and reverse speed ranges.

GEAR BACKLASH

Backlash is the clearance of "play" between two meshing gears. On normal gears (Fig. 23), the backlash is very small. However, gear wear, improper meshing of teeth, inadequate shaft support by bearings, or improper adjustment of the power train can cause excessive backlash.

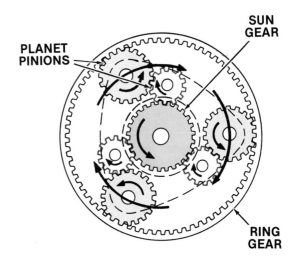

Fig. 23 — Backlash In Gears

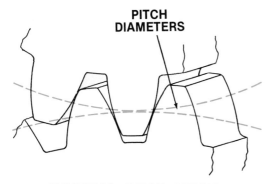

NORMAL GEAR MESH

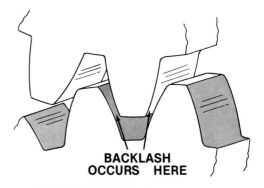

TOO MUCH BACKLASH

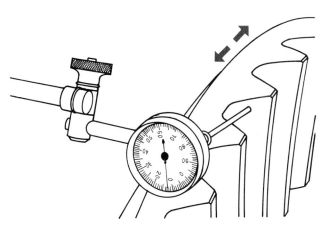

Fig. 24 — Checking Gear Backlash

Too much backlash can cause severe impact loading of gear teeth from sudden stops or gear reversal under load. It can also cause broken teeth and gear bouncing under load.

Too little backlash between gears causes overload wear on gear teeth, possibly premature gear failure, and excessive stress on shafts and bearings.

The technical service manual for most machines indicates the acceptable range of backlash in gears and describes procedures for adjusting backlash. To measure backlash, use a dial indicator (Fig. 24) to register the full rotary movement of the gear. Shims may be available for installation on either side of the gear to obtain the correct backlash setting.

GEAR ENDPLAY

A dial indicator may also be used to check endplay of gears and shafts (Fig. 25). Shims or adjusting nuts may be used to provide the desired range of

Fig. 25 — Checking Endplay Of Gears And Shafts

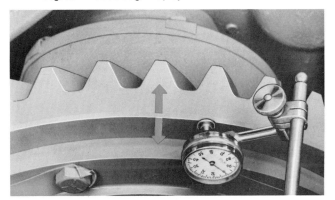

lateral movement of the gears or shaft for proper gear meshing.

GEAR WEAR

New gear teeth usually have slight imperfections, but these normally disappear during break in as the teeth are oiled and polished. After that, with proper lubrication, operation, and adjustment, the teeth should have a long service life.

However, when lack of lubrication or other factors cause gear failure, the cause can often be found by examining the broken parts. Major types of gear tooth wear and failures are shown in Fig. 26.

Normal wear appears as a polished surface which should extend the full length of the tooth from near the root (bottom) to the tip of the tooth. Gears properly manufactured and installed, well lubricated, and not overloaded will show this condition after long use.

Abrasive wear appears as surface injury from fine abrasive particles carried in the lubricant or imbedded in tooth surfaces. Common causes are metal particles worn from gears and abrasives such as sand and scale left in the gear case during casting.

Scratching is often found on heavily loaded gears operated at slow speeds. It's caused by metal particles (larger than abrasive particles) that flake off gears. It generally indicates the wrong gear design for the load. (Don't confuse with scoring.)

Overload wear is a depression along the length of the teeth caused when metal is removed by sliding pressure. Teeth are worn but are smooth. Continued wear results in excessive backlash and severe peening which may be misleading as to the real cause of the wear.

Rolling and peening are two problems. Rolling leaves a burr on the tooth edge caused by overloading and sliding of the teeth. Too little bearing support or too ductile a metal results in plastic flow of metal due to sliding pressure. However, peening results from excessive backlash and hammering of one tooth against another with tremendous impact. Lubricants are forced out and metal bears directly on metal.

Rippling leaves a wavy surface or "fish scales" on the teeth at right angles to the direction of slide. It may be caused by surface yielding due to heavy loads, vibration or "slip-stick" friction from lack of lubrication.

Scoring is caused by temperature rise and thinning or rupture of the lubricant film from overloading. Pressure and sliding action heats the gear and permits metal transfer from one tooth to the face of another. As the process continues, chunks of metal loosen and gouge teeth in the direction of sliding motion. The temperature rise here is slow and not as high as burning wear described below.

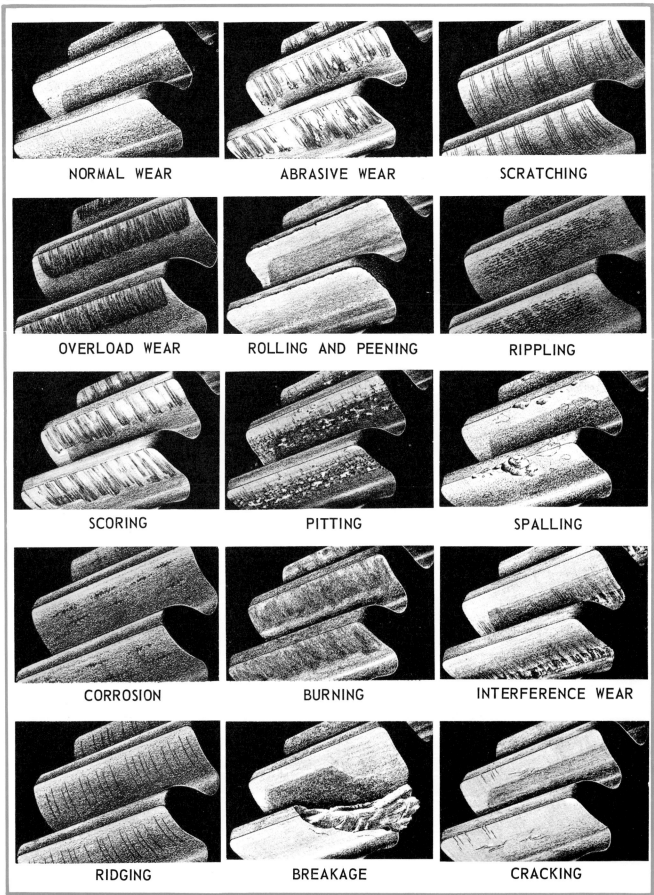

Fig. 26 — Gear Tooth Wear And Failures

Pitting is sometimes associated with thin oil film, possibly due to high oil temperatures. Very minute or micropitting can occur and would appear as a gray surface which may advance slowly to a visible pitted condition.

Spalling is common and starts with fine surface cracks and eventually results in large flakes or chips leaving the tooth surface. Improperly hardened teeth are most subject to spalling which may occur on only one or two teeth. But chips may then damage other teeth in the gear case.

Corrosion results in erosion of the tooth surfaces by acid. The acid is formed by moisture combining with lubricant impurities and contaminants from the air. Generally, surfaces become pitted, then uneven stresses lead to chipping and spalling.

Burning is usually caused by complete lubricant failure or lack of lubrication. Under high stress and sliding motion, friction develops rapid heating and the temperature limits of the metal are exceeded. Burned teeth are quite brittle and easily broken.

Interference wear can be caused by misaligned gears which place heavy contact on small areas. It can also be caused by mating gears of different tooth design. Different wear patterns may appear at tips and roots of gear teeth.

Ridging causes scratches near one end of a tooth, especially on a hypoid pinion gear. This can be caused by overloading, lack of lubrication, or improper heat treatment.

Breakage may result from many causes. Study teeth closely before assigning a cause. Breakage can be caused by high impact forces or defective manufacture. Examine the broken area closely. Fresh metal all over the break indicates damage from an impact overload. Fresh metal in the center with edges of the break dark or old looking means breakage from fatigue which started as a fine surface crack.

Cracking failures are usually caused by improper machining and faulty heat treatment during manufacture. Most heat treat cracks are extremely fine and don't show up until the gear has been used for some time.

For more information on diagnosis of gear and related component failure, refer to "Fundamentals of Service — Identification of Parts Failure."

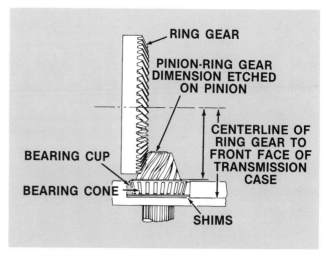

Fig. 27 — Adjusting Tooth Contact Between Ring Gear And Bevel Pinion

SERVICE AND MAINTENANCE OF GEAR DRIVES

Gears must mesh correctly and be properly aligned and lubricated for long service. Lubrication is covered later in this chapter under "Power Train Lubrication."

Gears which mesh too closely exert unnecessary force on gear teeth, shafts, and bearings. This causes rapid wear, early power train failure, and unnecessary expense as well as lost working time for the machine.

But portions of the gear teeth will be overloaded if teeth do not mesh fully or if gears are out of alignment. Be certain that gear shafts are parallel, or at the proper angle, and make shaft adjustments if necessary, usually by loosening shaft bearings and moving shafts slightly in relation to each other.

Proper meshing of bevel gears is often obtained by adding or removing shims below the bevel pinion (Fig. 27). In some ring gear-pinion sets the actual dimension is etched into the end of the pinion for easy reference during assembly or adjustment. As gear teeth are worn, more shims must be added to maintain correct tooth relationship.

To check gear tooth contact, coat several teeth on one gear with blueing and then turn the drive several revolutions (Fig. 28). By studying the pattern of blueing transferred to the mating gear teeth you can determine how well teeth match. If blueing is spread evenly over the teeth (A, Fig. 28), gears are well adjusted. However, if the pattern is too close to the tips of gear teeth (B, Fig. 28), or too near the base of the teeth (C, Fig. 28), the gears must be readjusted to provide correct mating.

It is usually advisable to replace both gears in a set if one is badly worn or damaged because teeth on a new gear will not mate properly with worn teeth on the old gear and could cause rapid wear of both gears. Some gears

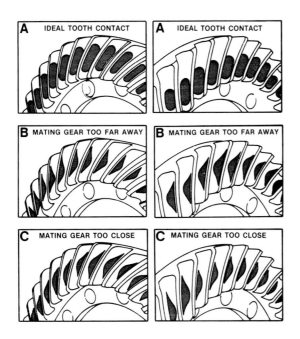

Fig. 28 — Checking Gear Tooth Contact Pattern

such as ring gears and pinions are sold only as matching sets to ensure proper mating.

BEARINGS

Bearings serve two major functions in power trains:

- **Reduce friction**

- **Support rotating shafts or parts**

Most power trains use antifriction bearings, those that provide a rolling contact between mating surfaces (Fig. 29). There are three main bearing types:

- **Ball**

- **Roller**

- **Needle**

Fig. 29 — Three Types Of Bearings Used In Power Trains

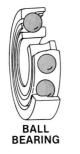

BALL   ROLLER   NEEDLE
BEARING   BEARING   BEARING

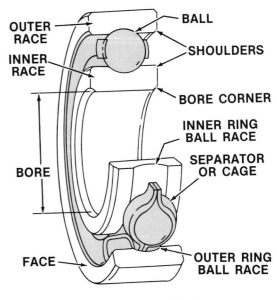

Fig. 30 — Basic Parts Of A Bearing

These bearings are made of:

1. Two hardened steel rings called races or cups (Fig. 30).

2. Balls, rollers, or needles which roll between the two races.

3. Optional separators to space the rolling elements around the bearing.

The outer or inner race is omitted in some bearings. Then the elements roll in direct contact with the shaft or other mounting (as in most needle bearings). When two races are used, one is normally pressed or fixed on a shaft or into a bore, while the other turns freely with the rolling elements. This is part of the antifriction design of these bearings.

Bearings may encounter one or two types of loading (Fig. 31):

- **Radial loads** — forces perpendicular to the axis of rotation

- **Thrust loads** — forces parallel to the axis of rotation

Replacement bearings must be of the same type as those originally installed to ensure proper performance, i.e., do not replace a thrust bearing with a radial load bearing and visa versa.

Ball Bearings

Ball bearings can support rotating parts or shafts for radial or thrust forces (Fig. 31). However, the shaft must align with the bore or the bearing will bind and fail.

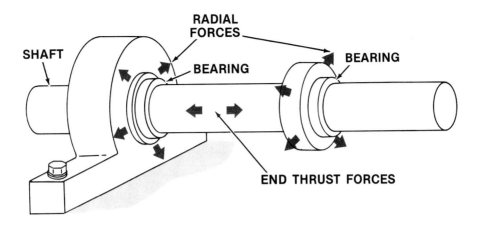

Fig. 31 — Load Forces Acting On Bearings

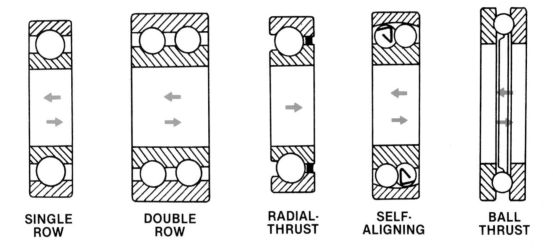

| SINGLE ROW | DOUBLE ROW | RADIAL-THRUST | SELF-ALIGNING | BALL THRUST |

Fig. 32 — Common Types Of Ball Bearings

Ball bearings designed to withstand radial and thrust forces (Fig. 32) are widely used. (Thrust forces are shown with small blue arrows.) Self-aligning ball bearings can compensate for small shaft angles in relation to the bearing mount.

Roller Bearings

Roller bearings are basically the same as ball bearings except the balls are replaced by rollers. In many bearings, the outer race can be removed without the rollers falling out. Depending on design, roller bearings can handle both radial and thrust forces. For heavy thrust and radial loads, tapered roller bearings are commonly used. Tapered bearings are also used where high pre-loads are required on shafts or gears to support thrust. Fig. 33 shows the common roller bearings.

Fig. 33 — Common Types of Roller Bearings

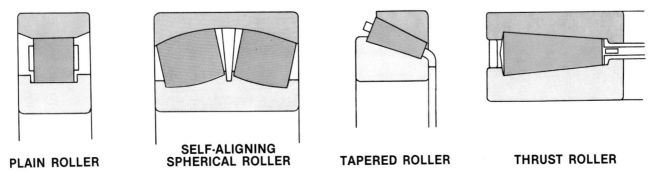

PLAIN ROLLER   SELF-ALIGNING SPHERICAL ROLLER   TAPERED ROLLER   THRUST ROLLER

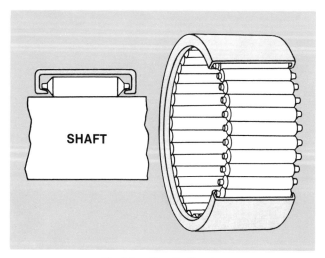

Fig. 34 — Needle Bearing

Needle Bearings

Needle bearings (Fig. 34) resemble roller bearings, but are thinner. Also, most needle bearings have no inner race and needles roll directly on the shaft. The rollers aren't separated but are tightly packed for added support of the shaft.

Needle bearings can support heavy radial loads but they cannot absorb thrust loads. They are commonly used in compact locations, such as inside a gear which must turn freely on a shaft or act as an idler. The length of the rollers and their tight packing provide good support and alignment. Planetary gears usually turn on needle bearings.

For more information on bearings refer to *FOS Bearings and Seals.* And when repairing equipment, if there is any doubt about bearing condition, replace the bearing.

Bearings may fail due to contamination, improper lubrication, improper installation, careless handling, distortion and misalignment, severe service, vibration, electric current and defects in bearing material. For detailed information on each of these types of failures, refer to "Fundamentals of Service — Identification of Parts Failures."

POWER TRAIN LUBRICATION

Proper lubrication is essential for satisfactory power train performance. Lubrication has three major functions:

- *Reduce friction and wear in power train parts.*
- *Protect parts from rust and corrosion.*
- *Help clean and cool moving parts.*

SELECTING PROPER LUBRICANTS

Power train lubricants range from standard motor oil and greases to specially-formulated oils for transmissions, hydraulic systems, and hydrostatic drives. The best lubricants are normally those recommended by the manufacturer in the machine operator's manual or the technical service manual.

Fig. 35 — Three Ways To Lubricate Open Gears

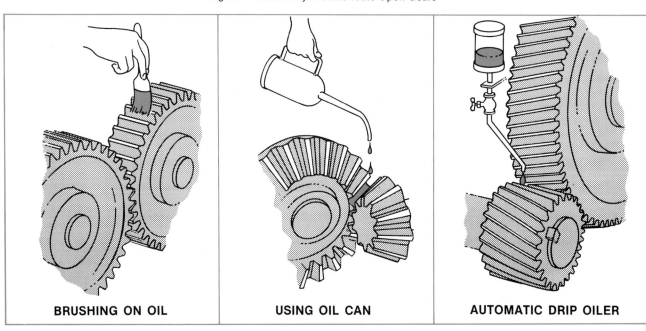

| BRUSHING ON OIL | USING OIL CAN | AUTOMATIC DRIP OILER |

Before substituting lubricants not recommended for a particular machine, it is advisable to contact the manufacturer to see if it will work and learn its affect on terms of the machine warranty.

For more information on lubricants refer to *FOS Fuels, Lubricants and Coolants*.

LUBRICATING OPEN GEARS AND CHAINS

Oil may be brushed or squirted on exposed gears (Fig. 35) or chains (Fig. 36) at intervals recommended by the manufacturer. The suggested intervals for oiling chains is commonly every eight hours with oil applied while the chain is warm to ensure good penetration.

Automatic drip or spray oilers may also be used but are generally restricted to use on stationary equipment.

Lubrication of some exposed parts is not recommended, particularly if the equipment is operated in adverse conditions such as extremely abrasive soil because lubricant may catch and hold dirt and increase wear more than would total lack of lubrication. Follow the manufacturer's instructions.

BEARING LUBRICATION

Many power trains have sealed bearings which do not require regular lubrication. However, some sealed bearings are equipped with grease fittings which should be lubricated at recommended intervals to maintain seal flexibility and flush out accumulated dirt and moisture. Do not over lubricate bearings. It could damage the seal and permit entry of dirt and water. Other bearings require greasing at intervals of 5 to 100 hours, depending on machine design, load, and operating conditions.

Always use the grease or oil recommended by the manufacturer.

CHANGING POWER TRAIN LUBRICANT

At the intervals suggested by the manufacturer, drain and refill with the recommended lubricant, all oil-bath gear and chain cases, transmissions, differentials, and hydraulic reservoirs.

If the unit has been disassembled for repair of broken or damaged parts, flush the inside with solvent to remove metal particles before refilling with new oil.

Many transmissions and hydraulic systems have an oil filter which must also be cleaned or changed at specified intervals. Note however that filter change intervals vary widely between manufacturers and machines. For instance, some recommend changing filters halfway between oil changes; others say to change filters and oil at the same time; while some specify filter changes

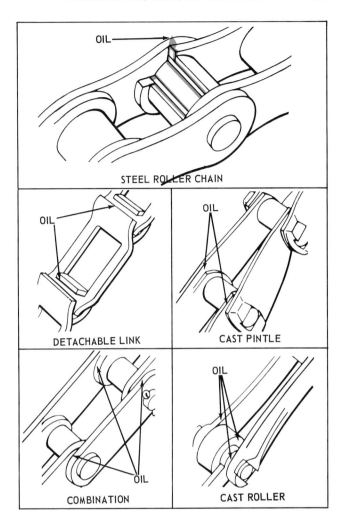

Fig. 36 — Lubrication Points On Chains

every other oil change. So, follow the operator's manual or technical service manual recommendations.

NOTE: Be sure to lubricate all bearings, chains and gears, or change oil at the suggested interval — or more frequently. Never operate equipment beyond the suggested intervals without providing recommended lubrication. If working under adverse conditions such as heavy dust, abrasive soil, deep mud or water, or during extremely hot or cold weather, service more often. Some engineers suggest cutting recommended lubricating intervals in half under severe operating conditions.

CHAPTER 2 REVIEW

1. Name three of the four sections in V-belt construction?

2. List four possible consequences of a misaligned belt drive.

3. (True or false) V-belts should ride on the sides of the sheaves, not on the bottom of the groove.

4. What can happen if belt drives are too loose or too tight?

5. List three types of drive chain and indicate which is used most on compact equipment.

6. List three advantages of chain drives compared to other drives.

7. List two reasons for maintaining proper chain drive tension.

8. (Choose one) A drive gear with 10 teeth will cause a driven gear with 20 teeth to turn (one-half, twice) as fast.

9. (True or false) It is generally impossible to tell by examination what may have caused a gear failure.

10. What is the function of bearings in a power train?

11. Name three common types of bearings.

12. Name three functions of power train lubrication.

13. (Choose one from each bracket) Power train lubrication intervals may be (reduced, extended) but should never be (reduced, extended).

CHAPTER 3

POWER TRAINS IN COMPACT EQUIPMENT

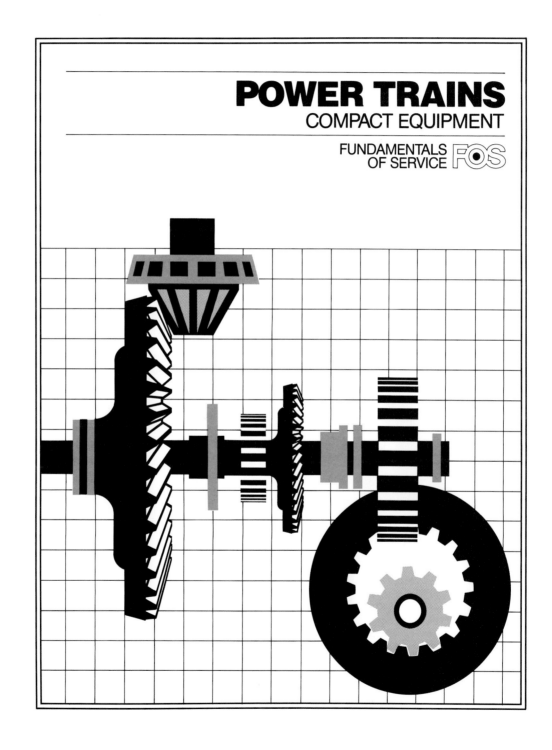

POWER TRAINS
COMPACT EQUIPMENT

FUNDAMENTALS
OF SERVICE FOS

SKILLS AND KNOWLEDGE

This chapter contains basic information that will help you gain the necessary subject knowledge required of a service technician. With application of this knowledge and hands-on practice, you should learn the following:

- **Power trains used in utility tractors.**

- **Power trains used in lawn and garden tractors.**

- **Power trains used in riding mowers.**

- **Power trains used in self-propelled walk-behind mowers.**

- **Power trains used in walk-behind rotary tillers.**

- **Power trains used in snow blowers.**

- **Power trains used in chain saws.**

- **Power trains used in powered hole diggers.**

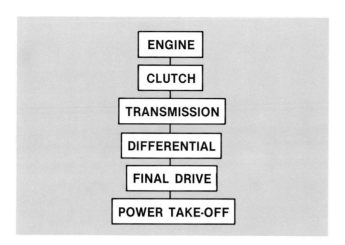

Fig. 1 — Machine Power Train Components Work Together

INTRODUCTION

Much of the diversity in compact products involves variations in the power train (Fig. 1).

You have seen how individual power train components transfer or control power flow, how they are serviced and maintained, and some problems that may be encountered in their application.

Now, let us look at how these components work together in equipment. We will show typical machines, the power train controls used, and the various combinations of power train components.

Machines to be illustrated include:

- **Utility tractors.**
- **Compact or skid-steer loaders.**
- **Lawn and garden tractors.**
- **Riding lawn mowers.**
- **Self-propelled walk-behind mowers.**
- **Walk-behind rotary tillers.**
- **Snow blowers.**
- **Chain saws.**
- **Powered hole diggers.**

UTILITY TRACTORS

Utility tractors (Fig. 2) perform in a wide variety of operating conditions. To meet these requirements, utility tractors are built with a broad combination of power train components (Fig. 3). Most of these components are controlled directly by the operator (Fig. 4).

Fig. 2 — Modern Utility Tractor

Fig. 3 — Utility Tractors Use Varieties Of Power Train Components

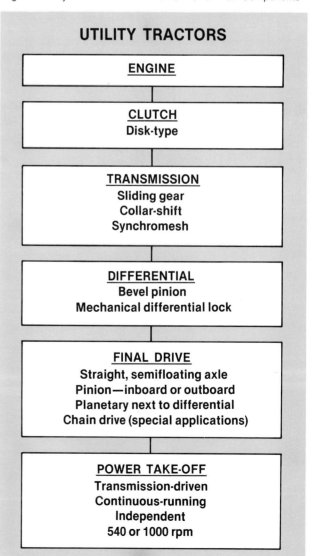

UTILITY TRACTORS

ENGINE

CLUTCH
Disk-type

TRANSMISSION
Sliding gear
Collar-shift
Synchromesh

DIFFERENTIAL
Bevel pinion
Mechanical differential lock

FINAL DRIVE
Straight, semifloating axle
Pinion—inboard or outboard
Planetary next to differential
Chain drive (special applications)

POWER TAKE-OFF
Transmission-driven
Continuous-running
Independent
540 or 1000 rpm

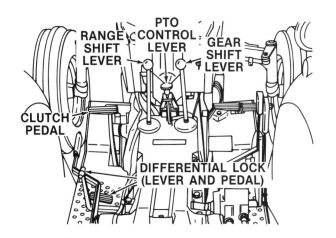

Fig. 4 — Operator Controls On A Utility Tractor

COMPACT LOADERS

Compact or skid-steer loaders (Fig. 5) are designed for maneuvering in very small spaces. To do this, a separate drive is provided to drive wheels on each side of the machine (Fig. 6). Power flow to each is controlled entirely by the operator (Fig. 7).

Applying power evenly to drive wheels on both sides causes the machine to move straight forward or reverse. But, if power is applied to only one side, either forward or reverse, the machine will make a wide circular turn. With the wheels driving forward on one side, and in reverse on the other side, the machine will pivot in its tracks. In effect, the operator supplies the differential action to the power train of these machines.

Fig. 5 — Skid-Steer Loaders Are Very Maneuverable

Fig. 6 — Power Flow In A Skid-Steer Loader

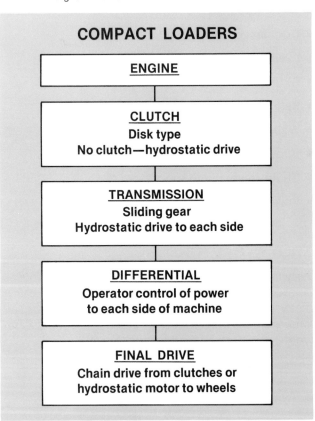

COMPACT LOADERS

ENGINE

CLUTCH
Disk type
No clutch—hydrostatic drive

TRANSMISSION
Sliding gear
Hydrostatic drive to each side

DIFFERENTIAL
Operator control of power
to each side of machine

FINAL DRIVE
Chain drive from clutches or
hydrostatic motor to wheels

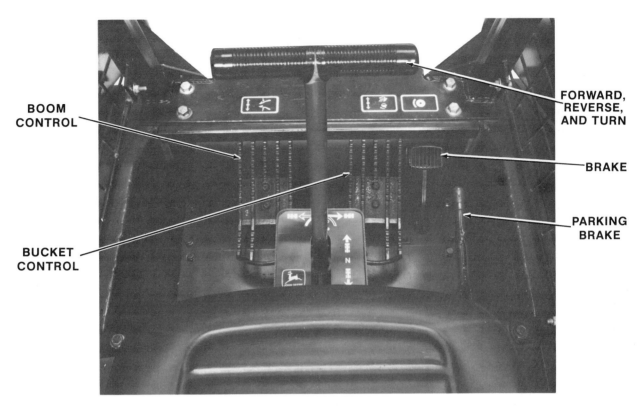

BOOM
CONTROL

BUCKET
CONTROL

FORWARD,
REVERSE,
AND TURN

BRAKE

PARKING
BRAKE

Fig. 7 — Power Train Controls For A Skid-Steer Loader

Fig. 8 — Modern Lawn And Garden Tractor

LAWN AND GARDEN TRACTORS

Lawn and garden tractors (Fig. 8) are available with many of the same power train components found on larger utility machines (Fig. 9), but they are not simply scaled down utility tractors. They use more belt drives and are designed for different applications. Transaxles (combined transmission and differential) are found on many of these tractors. The operator controls (Fig. 10) are similar to those on larger tractors.

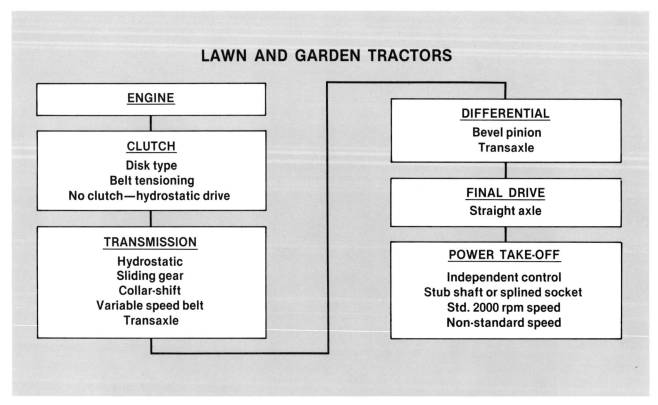

Fig. 9 — Lawn And Garden Tractors Use Many Power Train Combinations

Fig. 10 — Power Train Controls For Garden Tractor

Fig. 11 — Small Riding Mower

RIDING MOWERS

The power of riding lawn mowers ranges from about 5 to 20 horsepower (4 to 15 kw). Cutting widths vary from about 2 feet to 6 feet (0.6 to 1.8 m) or more. (This does not include heavy-duty units used to maintain parks, golf courses, and other large lawns, although some of the same power train features are used.)

Many small riding mowers (Fig. 11) look like miniature tractors. However, belt drives commonly transmit power to the transmission or transaxle (Fig. 12). Some have variable speed belt drives and most have from two to five forward speeds. Controls are all close to the seat (Fig. 13).

Some larger riding mowers have independent drives to wheels on each side of the machine (similar to skid-steer loaders — Fig. 5). One or two caster wheels support the rear of the machine (Fig. 14), but some models have two or even three drive wheels on each side for maximum traction and stability. A number of these machines use hydrostatic drives. Some have a separate control for drive motors on each side or for individual motors in each drive wheel (Fig. 15). Others use mechanical transmissions such as the cone and cup (see Chapter 5) or special clutching systems to direct power flow. Power train controls are very similar to those used on skid-steer loaders (Fig. 7).

RIDING MOWERS

ENGINE

CLUTCH
Belt tensioning
Disk type
No clutch—hydrostatic drive

TRANSMISSION
Sliding gear
Sliding key
Variable speed belt
Hydrostatic
Cone and cup

DIFFERENTIAL
Bevel pinion
Transaxle
Operator control of power to each side of machine

FINAL DRIVE
Straight axle
Hydrostatic motor in each wheel
Chain drive from hydrostatic motor to each wheel

Fig. 12 — Power Flow For Riding Mowers

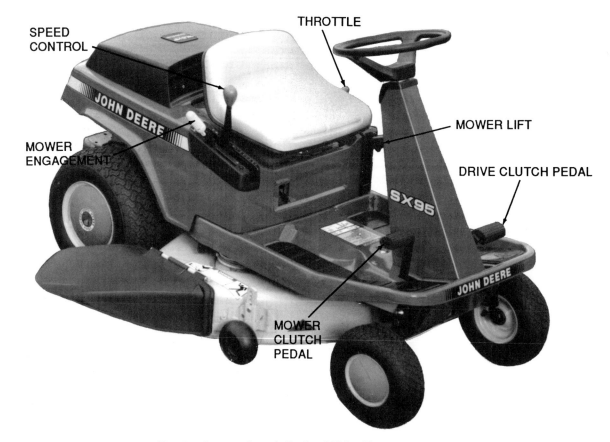

Fig. 13 — Operator Controls For Small Riding Mower

Fig. 14 — Mower With Independent Drive To Each Wheel

Fig. 15 — Separate Controls For Each Side Permit Mower To Turn Very Sharp

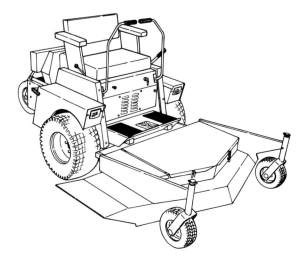

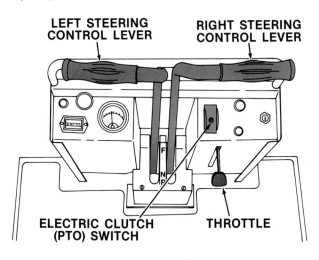

Fig. 16 — Self-Propelled Mowers May Have Front Or Rear-Wheel Drive

Fig. 17 — Walk-Behind Mowers Have Simple Power Train

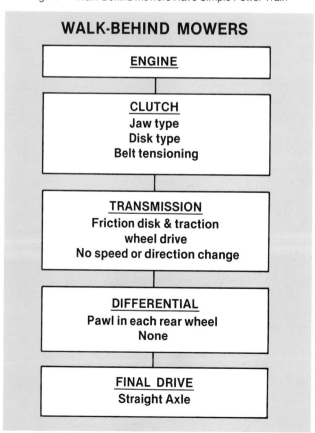

SELF-PROPELLED WALK-BEHIND MOWERS

Self-propelled walk-behind mowers (Fig. 16) may use either front or rear-wheel drive (Fig. 17). If front wheels are powered, no differential is required as the operator can simply push down on the mower handle and make a pivot turn without disengaging the power. A pawl in the hub of each rear wheel provides differential action for rear-drive models.

The clutch control (Fig. 16) is mounted on the mower handlebar for easy access.

Fig. 18 — Rotary Tiller With Worm Gear Drive To Tines

WALK-BEHIND ROTARY TILLERS

The majority of walk-behind tillers have no drive wheels and simply move forward (or backward) through rotation of the tilling tines (Fig. 18). However, some tillers have separate drive wheels and may have several travel and tine speeds. Tiller power trains are simple and rugged (Fig. 19) and include various combinations of belts, drive chains, worm gears and on some machines, a sliding gear transmission.

Fig. 19 — Rotary Tillers Have Several Types Of Power Trains

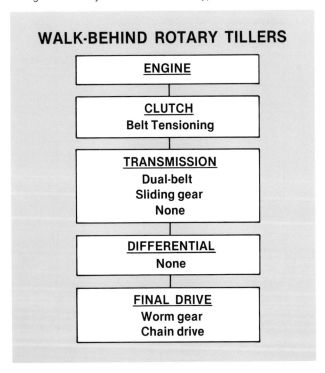

SNOW BLOWERS

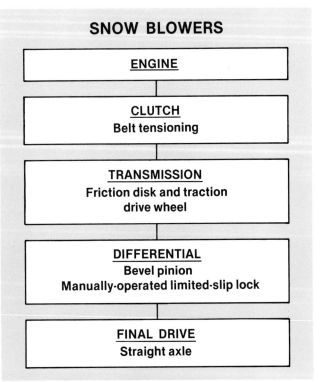

Fig. 20 — Power Flow In A Snow Blower

SNOW BLOWERS

Power flow to snow blowers (Fig. 20) must be smooth and dependable (Fig. 21) — even at very low temperatures. The machines must also have adequate traction for maneuvering under very adverse conditions. And they must provide safe working conditions for the operator, even when operating in heavy snow or ice. Therefore, most current snow blowers are designed to automatically stop the engine or stop power flow to the auger if the operator releases safety levers without first disengaging the auger drive (Fig. 21).

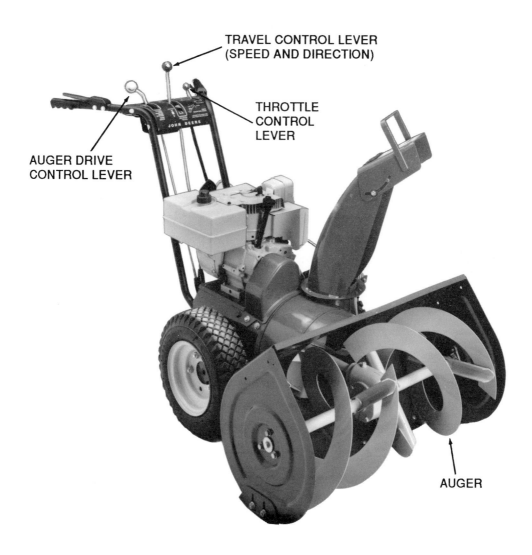

Fig. 21 — Snow Blowers Operate In Very Adverse Conditions

Fig. 22 — Centrifugal Clutch Transmits Power To Saw Chain

CHAIN SAWS

Chain saws (Fig. 22) use centrifugal expanding shoe clutches which are engaged and disengaged by changing engine speed. As the speed increases, shoes in the rotating clutch move outward against a driven member and power flows to the saw chain (Fig. 23). Releasing the throttle control permits clutch springs to retract the shoes and break power flow.

Fig. 23 — Chain Saw Power Flow

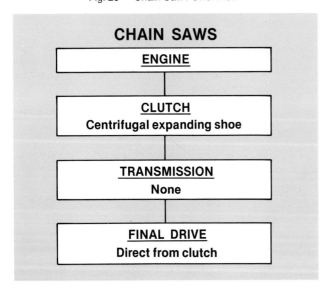

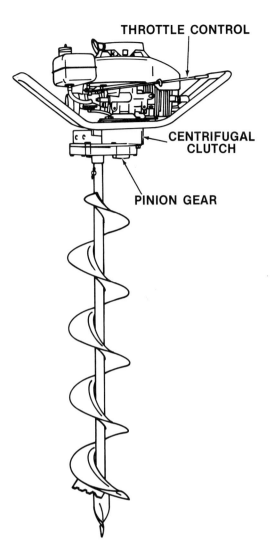

THROTTLE CONTROL

CENTRIFUGAL CLUTCH

PINION GEAR

Fig. 24 — Powered Hole Digger Uses Expanding Shoe Clutch

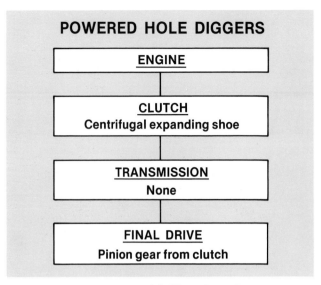

POWERED HOLE DIGGERS

ENGINE

CLUTCH
Centrifugal expanding shoe

TRANSMISSION
None

FINAL DRIVE
Pinion gear from clutch

Fig. 25 — Powered Hole Digger Power Flow

POWERED HOLE DIGGERS

Powered hole diggers (Fig. 24) are used to dig holes for setting posts or poles, for planting shrubs and small trees, and to cut holes in ice. Most diggers use a centrifugal expanding shoe clutch (similar to those used in chain saws) and a pinion gear to power the auger or drill (Fig. 25). Increasing engine speed engages the drive — reducing speed stops power flow.

CHAPTER 3 REVIEW

1. Which machine can turn in its own tracks and how is this done?

2. (True or false) Most riding mowers have belt drives.

3. (True or false) Self-propelled walk-behind mowers use only front wheel drive.

4. (True or false) A differential is not required in front wheel drive walk-behind mowers.

5. Which walk-behind implement has no drive wheels and how is it pulled forward?

6. Which implements use a centrifugal expanding shoe clutch?

CHAPTER 4

CLUTCHES

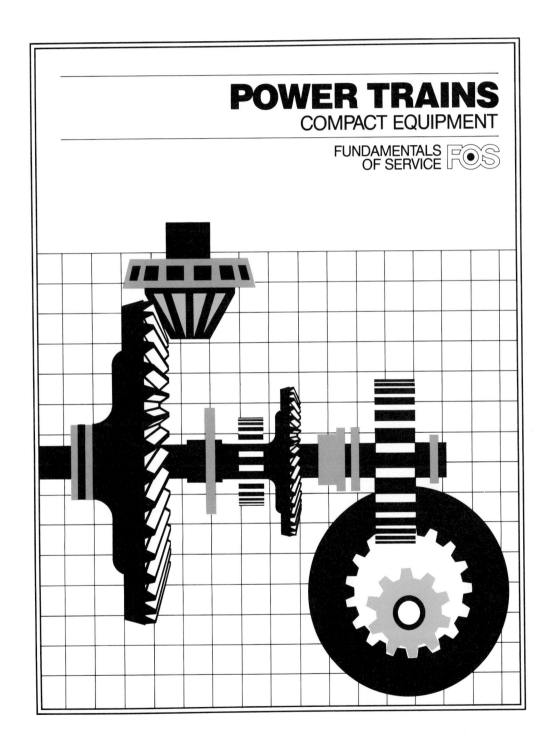

POWER TRAINS
COMPACT EQUIPMENT

FUNDAMENTALS OF SERVICE FOS

SKILLS AND KNOWLEDGE

This chapter contains basic information that will help you gain the necessary subject knowledge required of a service technician. With application of this knowledge and hands-on practice, you should learn the following:

• **Types of clutches used in compact equipment and how they operate.**

• **Procedures for servicing and maintaining these different clutches.**

• **Service and maintenance procedures for mechanical clutch controls.**

• **How to analyze and troubleshoot the eight basic troubles of disk and plate clutches.**

INTRODUCTION

The clutch connects and disconnects power between the engine and the transmission or working component.

In some machines the clutch provides a "safety valve" so if the power train is overloaded, the clutch slips to avoid more serious damage to other power train parts.

Chapter 1 explained the need for and operation of a clutch in the power train and described the basic types of clutches available. Now we will see in more detail how these clutches operate and how they are serviced and maintained.

TYPES OF CLUTCHES

The clutches to be discussed in this chapter include:

● *Belt tensioning clutches.*

● *Jaw clutches.*

● *Friction clutches.*

    a. *Disk and plate*

    b. *Band*

    c. *Expanding shoe*

    d. *Cone*

● *Electromagnetic clutches.*

● *Centrifugal, overrunning, and load-sensitive clutches.*

BELT TENSIONING CLUTCHES

When a belt tensioning clutch is engaged, a drive belt is tightened so that power flows from the drive sheave (usually on the engine) to a driven sheave on the transmission (Fig. 1), or a working component such as a mower blade. Disengaging the clutch releases the belt tension, and the drive sheave is free to rotate without driving the other parts of the power train.

Belt tensioning clutches on the ground drive system of riding mowers and lawn and garden tractors are usually controlled by a foot-operated clutch pedal. Belt tensioning clutches for mower drives or power take-offs on these machines are usually hand operated. In most ground drive systems, a spring-loaded idler holds the belt tight. Depressing the clutch pedal releases the belt tension thus stopping power flow to permit shifting transmission gears.

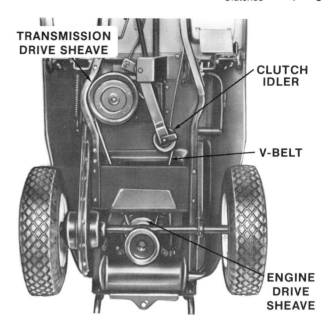

Fig. 1 — Belt Tensioning Clutch

A hand clutch may be used to shift the idler-tightener "over-center" where it is held in place by belt tension and linkages (Fig. 2). Other machines require the operator to hold the clutch control lever in the engaged position. Releasing the hand lever disengages the clutch, a good safety factor on walk-behind mowers and snowblowers. This system is sometimes called a "deadman" clutch.

Fig. 2 — Over-Center Belt Tensioning Clutch

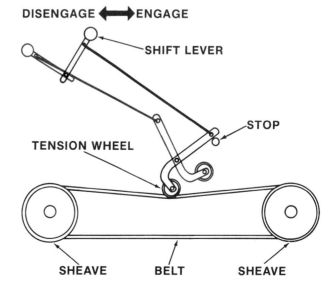

SERVICE AND MAINTENANCE OF BELT TENSIONING CLUTCHES

If belt tensioning clutches start slipping, adjust the linkage between the clutch control and the idler to increase tension on the drive belt when the clutch is engaged (Fig. 3). If the belt continues to slip after the adjustment reaches the maximum, see if it is possible to reposition the engine or the countershaft to remove additional slack from the belt. Be sure to recheck drive alignment after moving these components.

If the components cannot be repositioned, install a new belt specified for that machine. Other belts may fit but will not likely meet specifications for strength required for the application.

Check alignment of the clutch-idler sheave with the drive and driven sheaves regularly. Adjust alignment if necessary (see Chapter 2). Also inspect the idler linkage for excessive freeplay or wear that could throw the belt out of alignment when the clutch is engaged. Inspect the idler bearing and be sure the idler sheave turns freely without wobbling or vibrating. Replace the idler if necessary.

Follow the manufacturer's instructions for lubricating the idler. Many drives have sealed bearings which require no lubrication, but always check to be sure. Just avoid overgreasing which may get lubricant on the belt and cause slippage and belt deterioration (Fig. 4).

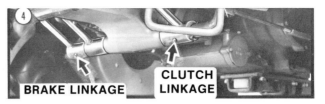

Fig. 4 — Lubricating Clutch Linkage

JAW CLUTCHES

Jaw clutches are mainly used for low-power, slow-speed drives because of their abrupt engagement. When the jaws slide together, power immediately flows through the drive. There is no slippage.

The clutch jaw (Fig. 5) slides on a spline to connect and disconnect power flow. The shifter fork is spring-loaded to the engaged position. Moving the clutch control lever to the disengaged position pivots the shifter arm and disengages the jaw clutch.

SERVICE AND MAINTENANCE OF JAW CLUTCHES

If a jaw clutch malfunctions, first check for slippage of the V-belt (Fig. 5) driving the clutch gearbox. If the belt is satisfactory, readjust the clutch control linkage to be certain the shifter arm is free to move between the fully engaged and disengaged positions when the control lever is shifted. Replace the engaging spring (Fig. 5) if it is weak or broken. If the problem is inside the gearbox:

● *Inspect clutch jaws for wear or damage.*

● *Check the worm gear for wear or damage.*

● *Be sure the gearbox is filled with the proper lubricant.*

FRICTION CLUTCHES

Several friction clutches were described in Chapter 1. We will now take a closer look at the operation and care of:

● **Disk clutches.**

● **Plate clutches.**

● **Dry clutches.**

● **Wet clutches.**

● **Band clutches.**

● **Expanding shoe clutches.**

● **Expanding cone clutches.**

Fig. 3 — Adjusting Clutch Linkage

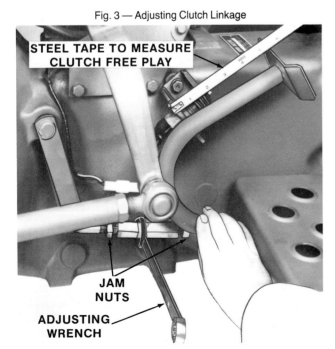

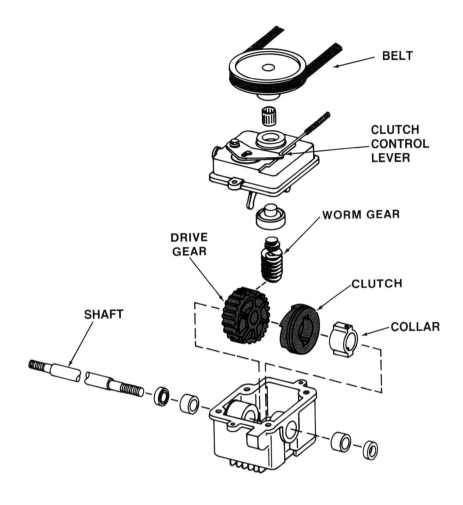

Fig. 5 — Jaw Clutch

Disk and Plate Clutches

Two major types of disk and plate clutches are used:

- *Dry — Operates dry and is cooled by air.*

- *Wet — Operates in oil spray or bath and is cooled by oil.*

Note that materials used in these two types of clutches are quite different. Oil on a dry clutch can cause clutch slippage and require replacement. But the disks and plates in wet clutches must be coated with oil to operate properly. Wet clutches usually require specific types of oil for each clutch .

Other than the coolant used, operation of wet and dry clutches is essentially the same. However, wet clutches are generally found on larger equipment and are not often used on compact products. Therefore, they will not be covered in detail.

Dry Disk Clutch: A typical dry clutch (Fig. 6) has a driven disk with friction facing (red areas) molded to both sides. The clutch pressure plate housing is bolted directly to the engine flywheel, and the driven disk is splined to the transmission input shaft.

Depressing the clutch pedal moves the clutch release bearing against the operating levers (Fig. 6). As the levers pivot, the clutch pressure plate is pulled away from the disk, compressing the clutch springs. This separates the driven disk from the pressure plate so the flow of power between the flywheel and the transmission input shaft is stopped.

Releasing the clutch pedal permits the clutch springs to push the pressure plate against the clutch disk so the flow of power is reconnected between the engine and transmission.

A typical dry clutch disk and pressure plate are shown in Fig. 7.

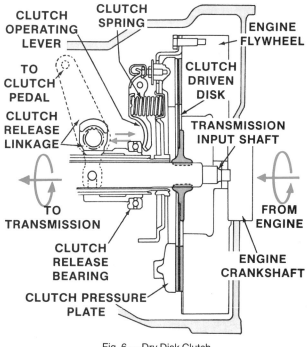

Fig. 6 — Dry Disk Clutch

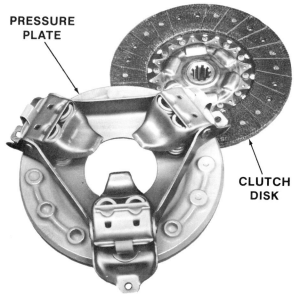

Fig. 7 — Dry Clutch Disk And Pressure Plate

Some heavy-duty clutches have several plates and disks (Fig. 8) to transfer power.

Clutch disks may be lined with an organic facing material to provide a nonslip surface. Or linings may be made of a ceramic-metallic mixture which is oven-baked and quite durable. The lining must be long-wearing, heat-resistant, and able to resist slippage under heavy loads. The linings may be bonded to the disk or attached with countersunk rivets. To prevent damaging the flywheel or pressure plate, linings must be replaced before they are worn enough to cause metal-to-metal contact. The phrase "worn down to the rivets" is often used to describe a worn clutch.

Fig. 8 — Parts Of A Heavy-Duty Disk Clutch

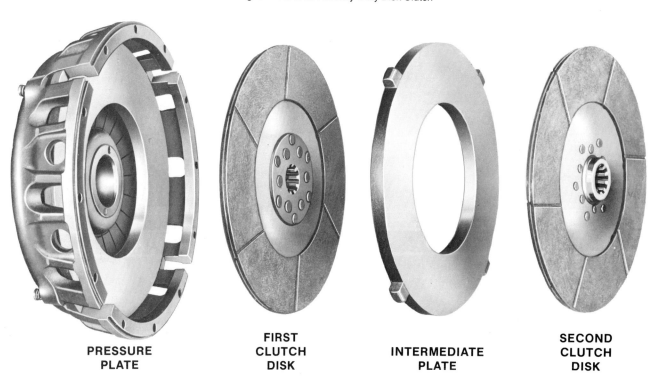

PRESSURE PLATE   FIRST CLUTCH DISK   INTERMEDIATE PLATE   SECOND CLUTCH DISK

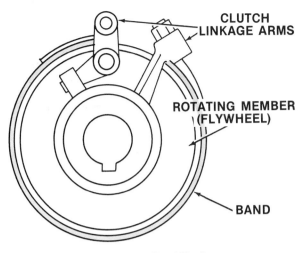

Fig. 9 — Band Clutch

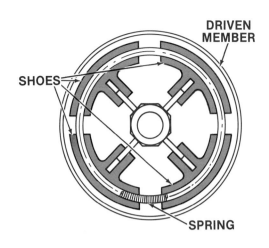

Fig. 11 — Centrifugal Expanding Shoe Clutch

Band Clutches: In band clutches, a band or belt is squeezed around a rotating flywheel or drum (Fig. 9). Band linkage is coupled to the driven member of the power train. These clutches are bulky and seldom used on compact equipment and will not be discussed in detail.

Expanding Shoe Clutches: Mechanical expanding shoe clutches have a release bearing (Fig. 10) which is slid in or out on the driven shaft by the clutch control pedal or lever. Pivoting linkage attached to the release bearing forces clutch shoes outward against a drum (outer member) or retracts shoes to disengage the clutch. Mechanical expanding shoe clutches are not commonly found on compact products and will not be discussed in further detail.

Fig. 10 — Mechanical Expanding Shoe Clutch

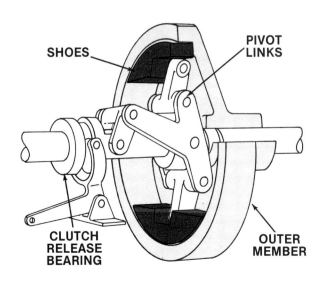

Centrifugal expanding shoe clutches however are much more compact, lighter, and are commonly found in compact products. As the drive shaft in a centrifugal clutch rotates, centrifugal force causes clutch shoes to move outward (Fig. 11) until they contact the outer member. As drive speed increases, the shoes are pressed harder against the outer member rotating it at the same speed as the drive unit. When speed decreases, springs pull the shoes away from the outer member and power flow is stopped.

These clutches are commonly found on chain saws and other hand-carried power equipment because of their small size, low weight, and convenience. Since most such machines are operated at full engine throttle, the operator must activate only one control — the throttle — to engage or disengage the drive and regulate machine operation.

Cone Clutches: Cone clutches are engaged and disengaged by sliding a throwout bearing in or out on the driven shaft. When the cone members are forced together (Fig. 12), the clutch lining on the driven member engages the machined surface of the drive member and power flows through the system. The drive member may be the engine flywheel or a specially-shaped V-belt sheave mounted on the engine crankshaft or elsewhere in the power train.

Cone clutches are compact and provide control of auxiliary drives such as power take-offs and mowers on lawn and garden tractors.

SERVICE, MAINTENANCE, AND REPAIR OF FRICTION CLUTCHES

All friction clutches have one thing in common — special surfaces, called linings, to provide friction between the

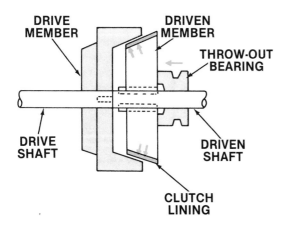

Fig. 12 — Cone Clutch

driving and driven members of the clutch. These linings wear out, but their life depends on the following factors:

- *Frequency of clutch use: Start-and-stop operation increases clutch wear.*

- *Clutch size in relation to power and load.*

- *Lining material.*

- *Operator control: "Popping" or jerking the clutch wears the lining fast; so does slipping the clutch during engagement; and "riding" the clutch (keeping the operator's foot resting lightly on the pedal during operation).*

- *Overlubricating the release bearing or linkage makes a clutch slip, grab, and wear out before it should.*

Fig. 13 — Provide Safe Supports For Engine And Clutch Housing

Fig. 14 — Disassemble Clutch Shaft And Release Mechanism

Clutch linings must be replaced before they are worn enough to damage mating surfaces. If a clutch pressure plate or flywheel has become scored due to excessive lining wear, they must be replaced or machined smooth before installing new clutch linings or the scored surface will soon wear out the new linings.

Let's see what is involved in making clutch repairs.

Repairing A Disk Clutch

To remove clutch, separate the clutch housing from the engine as instructed in the technical service manual. Be sure to provide sturdy supports for each portion of the machine to prevent accidents (Fig. 13). Less common band, cone, and expanding shoe clutches require different procedures. Check the appropriate technical manual for maintaining them.

To inspect the clutch shaft and release mechanism, remove bearing release springs (Fig. 14), the clutch release bearing, and the clutch shaft. Disconnect clutch pedal linkage from the outside and remove the clutch release yoke and linkage from the housing.

Inspect the clutch shaft (Fig. 15) and replace it if splines are worn or damaged or if shaft diameter at the release bearing or at the pilot bushing is worn beyond the acceptable tolerance specified in the technical service manual.

Inspect the release bearing and replace it if it is loose on the sleeve, if it appears burned, or if it is worn beyond the specified tolerances. Also replace the clutch shaft and release bearing if the clutch and bearing fail to slide

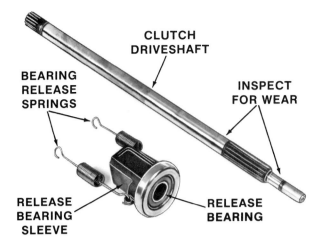

Fig. 15 — Inspect Clutch Drive Shaft And Release Bearing For Wear Or Damage

freely between the engaged and disengaged positions. Inspect release springs and measure spring length at the specified force. Replace springs if they fail to meet these requirements.

Inspect the clutch yoke, yoke arm shaft, and pedal assembly. Replace worn or damaged parts.

Now turn to the clutch itself. Before removing the clutch from the engine, check the release plate height (Fig. 16) and write down the measurement for use in reassembly. Remove the capscrews holding the clutch cover to the flywheel and remove the cover assembly. Do not drop the clutch drive disk.

 CAUTION: Use proper equipment and methods as prescribed in the technical service manual when disassembling a clutch pressure plate. Spring pressure must be kept under control and released evenly to avoid injury.

IMPORTANT: Do not remove the nuts holding the clutch release levers to the clutch cover as they are set at the factory and must not be disturbed.

On some machines, if the clutch cover assembly and pressure plate is damaged or worn, the entire assembly must be replaced.

Check the flywheel pilot bushing (or bearing) for wear or damage (Fig. 17) and inspect the flywheel surface for scratches and cracks. Clean any oil or rust off the flywheel surface with a light abrasive. Replace the pilot bushing if it fails to meet the specified tolerances.

Measure clutch plate thickness (Fig. 18) and replace the plate if it is worn too thin. Also replace a clutch plate damaged by heat or distortion and dry clutch plates which have been exposed to grease or oil.

While the clutch is disassembled, clean all the parts and compare dimensions of parts to the tolerances in the technical manual. Do not get oil or solvents on friction surfaces. Replace damaged and worn parts.

Check the flywheel friction surface for roughness and uneven wear. Use a straightedge and feeler gauge at several points around the flywheel (Fig. 19). Use the same procedure to measure flatness of the pressure

Fig. 16 — Measure Release Plate Height

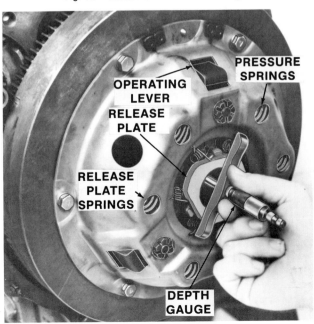

Fig. 17 — Inspect Pilot Bearing And Flywheel Drive Surface

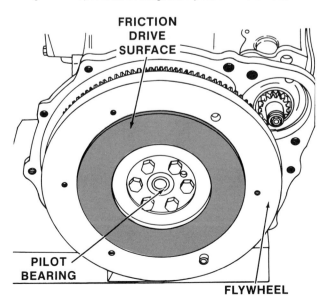

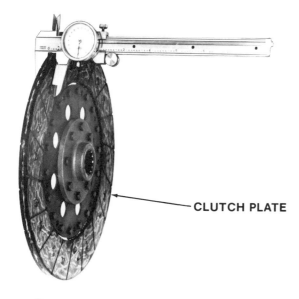

Fig. 18 — Measure Clutch Plate Thickness For Wear

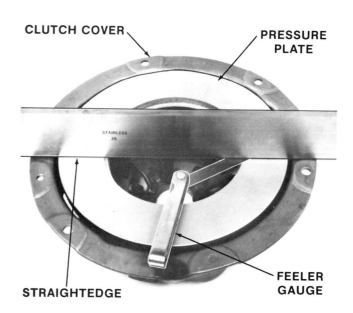

Fig. 20 — Measuring Clutch Pressure Plate Flatness

plate (Fig. 20). If either surface is not within the specified flatness, the part must be machined flat or replaced. The flywheel must also be replaced if it is damaged by excessive heat cracks.

After all clutch parts and related items have been cleaned and inspected and replacement parts obtained for worn or damaged items, reassemble the clutch. Remember that even seemingly minor differences in machine design can require entirely different procedures for assembly and adjustment. In addition, improper assembly can cause clutch malfunction and premature failure. Therefore, it is vitally important that technicians follow the steps specified in the technical service manual when repairing or replacing the clutch.

Fig. 19 — Measuring Flywheel Friction Surface Flatness

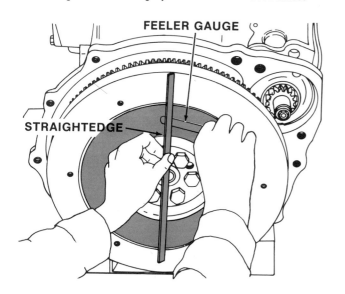

This helps avoid damage to parts, mistakes in reassembly, or personal injury. Be sure to tighten all nuts and capscrews to the specified torque and use locking devices where indicated to reduce the possibility of parts working loose during machine operation.

ELECTROMAGNETIC CLUTCHES

Electromagnetic clutches are essentially friction clutches in terms of transmitting power. But electromagnets engage the clutch and hold it in the operating position instead of springs or mechanical linkage. When the electric circuit to the magnets is broken (clutch pedal or lever disengaged), the clutch is released and power flow stops.

Unlike manually-engaged friction clutches, which can be slipped or jerked into engagement by the operator, electric clutch engagement is smooth and uniform each time. The armature (Fig. 21) is keyed to the engine crankshaft and therefore turns whenever the engine is running. When an electric current energizes the field windings, the resulting magnetic field pulls the spring-loaded armature against the clutch rotor which (in Fig. 21) carries drive sheaves for an auxiliary drive. The rotor and sheaves are mounted on ball bearings so the drive shaft can rotate freely without turning the rotor when the clutch is disengaged. The field assembly is also mounted on ball bearings so it can remain stationary while the rotor turns. There is no spring-loaded pressure plate in an electric clutch, and no direct physical contact is made between the magnet (field windings), the rotor, and the armature.

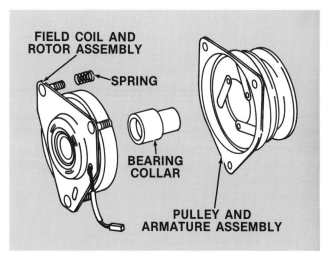

Fig. 21 — Electromagnetic Clutch Parts

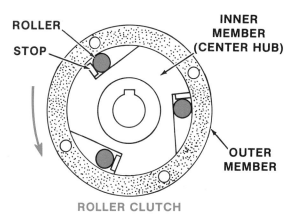

Fig. 22 — Roller Overrunning Clutch

SERVICE AND MAINTENANCE OF ELECTROMAGNETIC CLUTCHES

If an electric clutch malfunctions, first check for current flow to the field windings when the clutch control is "engaged." If there is no current, check for loose wiring, a faulty switch, or defective thermal relay in the circuit. Follow other steps prescribed in the machine technical service manual.

If there is current but no clutch action, remove field assembly from the clutch (with wires attached) and activate the switch. Hold a piece of iron close to the coil. If it is not attracted by the magnet, replace the field assembly.

Inspect the field, rotor, and armature for damage. If the field shows black or burned areas, replace it. Also check the armature and face side of the rotor for blue or black discoloration of steel which indicates slippage. If slippage has overheated the rotor enough, it may have burned the field, which must then be replaced too.

Check the rotor for signs of rubbing against the field. Inspect bearings for wear or damage. Check specifications for the proper air gap between the rotor and armature when the clutch is disengaged. Install or remove shims between the rotor and armature as indicated in service instructions to obtain the proper gap.

Some electromagnetic clutches must be replaced as a unit if they malfunction. Others have the armature, rotor, and bearing in one assembly and the field and bearing in another. Other electromagnetic clutches, usually those for large equipment, may be serviced by replacing individual parts.

CENTRIFUGAL, OVERRUNNING, AND LOAD-SENSITIVE CLUTCHES

There are several types of centrifugal and overrunning clutches available, including the centrifugal expanding shoe clutches discussed earlier. Three other types include:

- **Roller.**
- **Cam or sprag.**
- **Spring.**

Roller Clutch: When the center hub of a roller clutch (Fig. 22) is rotated, the rollers roll outward along the ramps in the hub. As the speed increases, the rollers wedge against the outer member and cause it to turn with the center drive hub. Slowing the center hub lets the rollers return and the outer member is free to coast to a stop or "overrun" the drive.

Cam or Sprag Clutch: The sprag clutch works much like the roller clutch except the rollers are replaced by a row of cams or sprags. As the inner drive member is rotated, the cams or sprags are tilted as they try to follow the rotating surface (Fig. 23). This wedges the outer por-

Fig. 23 — Sprag Or Cam Overrunning Clutch

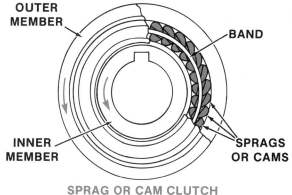

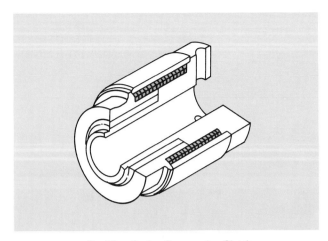

Fig. 24 — Spring Overrunning Clutch

tion of the sprags against the outer member, forcing it to turn with the center hub.

If the center hub slows down, the cams lean the other way and disengage the driving action. The band of material through the center of the sprags prevents them from lying flat.

Spring Overrunning Clutch: For some smaller applications, a spring clutch is used. A coil spring wrapped tightly around a shaft or collar (Fig. 24) tightens on the shaft when the drive is rotated in one direction. When power is removed from the drive, or the drive is reversed, the spring loosens and does not drive.

SERVICE AND MAINTENANCE OF OVERRUNNING CLUTCHES

Failure of a centrifugal or overrunning clutch to disengage or freewheel when the drive member slows or is reversed defeats the purpose of the clutch.

These clutches must be kept free of rust, corrosion, and dirt which interfere with roller or sprag action. But, excessive lubrication, heavy, tacky grease, or grease thickened by age or low temperatures can also interfere with clutch action.

Badly worn rollers or cams may fail to engage the outer member or slip under heavy loads. If the drive or driven members become scored from repeated engagement in the same spot, the clutch may fail to properly disengage. So, any worn or damaged parts must be replaced. Always replace rollers or sprags as a set, or replace the entire clutch. Otherwise the newer parts would be engaged and carry the entire load before older, worn-down parts are able to contact the outer member.

Refer to the technical service manual for specific clutch service and maintenance instructions and tolerances.

LOAD SENSITIVE CLUTCHES

The load sensitive clutches commonly found on compact equipment are a variation of the variable-speed belt drives described in Chapter 1. When the engine is idling, the drive is disengaged. As engine speed increases the drive engages. If the load is minimal, the variable speed sheaves quickly shift to maximum speed in relation to the engine speed. However, if the load is heavier, the sheaves assume a higher-torque, lower-speed configuration providing more power at a slower travel or working speed.

CLUTCH OPERATING MECHANISMS

Clutches are normally operated in one of four ways:

● **Mechanically**

● **Hydraulically**

● **Electrically**

● **Pneumatically**

We will discuss the mechanical clutches because electric clutches were discussed earlier in this chapter, and hydraulic clutches, as was pointed out in Chapter 1, are commonly found only on large tractors and equipment and are of little concern in terms of compact equipment. Likewise, pneumatic clutches which use air to operate are not discussed because they are not used on compact equipment.

MECHANICAL CLUTCH CONTROL

Mechanical clutch controls — pedals, levers, springs, and linkages — are used to manually or mechanically engage and disengage clutches. Mechanical controls may be used on jaw and belt-tightening clutches as well as friction clutches. Three basic types of controls are used:

● **Standard**

● **Over-Center**

● **"Dead-Man"**

Standard linkage is the most common; the clutch remains engaged unless a pedal or lever is moved and held in the disengaged position.

Over-center linkage provides a mechanical advantage to lock the clutch in the engaged or disengaged position. It is frequently used with belt tensioning clutches and may be found on mechanical, expanding shoe clutches.

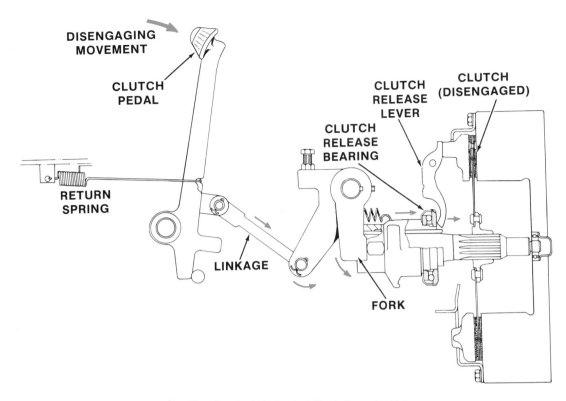

Fig. 25 — Standard Mechanical Clutch Operating Linkage

"Dead-man" clutch controls must be held in the engaged position by the operator. In some cases the control lever may be held in place by a thumb button which is released by squeezing the handle.

Standard linkage: Motion of the clutch pedal (or possibly a lever) is translated through linkage to the clutch release bearing fork (Fig. 25). This pushes the release bearing against the clutch release levers, releasing or disengaging the clutch. Springs may be used to aid in disengaging the clutch, as well as to engage the clutch whenever the pedal is released.

Over-center linkage: Moving clutch linkage "over-center" locks it in place and prevents disengagement unless the control is returned to its original position.

Over-center linkage may be used on belt-tensioning clutches, expanding shoe clutches, or with the hand-operated clutches found on some older tractors.

The over-center belt clutch (Fig. 26) is engaged by pushing forward on the control lever until the idler pulley passes the lowest point in its arc of travel and relaxes some of the pressure on the belt. A stop on the lever or control arm prevents further rotation of the idler. Thus the idler is held in place by belt tension. Pulling back on the control lever rotates the idler free of the belt.

Dead-man controls: These are commonly used on walk-behind equipment to automatically disengage power flow if the operator releases the clutch control (Fig. 27). This helps prevent accidental injury if the operator should fall or stumble while running a snowblower or mowing on a slope with a self-propelled mower. It also reduces the chances of injury by stopping equipment automatically in case the operator is tempted to

Fig. 26 — Over-Center Belt Tensioning Clutch

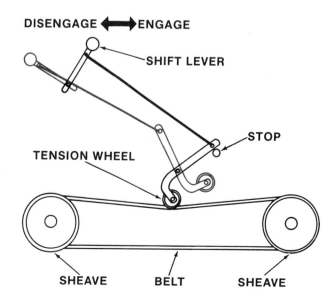

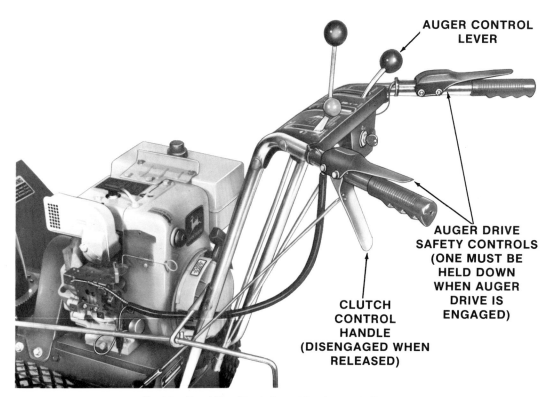

Fig. 27 — Dead-Man Clutch Control For Automatic Disengagement

leave the machine running while unplugging, lubricating, or adjusting some part of the machine.

A quick squeeze on the clutch control releases the thumb button used on some equipment to reduce operator fatigue. The drive is thus quickly and easily disengaged for sudden stops and safety of the operator and bystanders.

Dead-man controls are installed for the operator's protection. Owners and operators should be strongly discouraged before attempting to bypass, disconnect or circumvent their operation in attempts to "save time" or make the machine "easier to operate."

As an added safety feature on many machines, the clutch control must be in the "disengaged" position before the engine will start, whether the transmission is in gear or in neutral. For more information on "safety start" controls refer to *FOS Electrical Systems* and the technical service manual for the particular machine.

SERVICE AND MAINTENANCE OF MECHANICAL CLUTCH CONTROLS

Many clutch malfunctions and failures can be traced to improper clutch adjustment, worn bushings, bent rods, broken springs, worn bearings, or damaged cotter pins. These problems can require excessive force to operate the clutch and may prevent complete engagement or disengagement. This results in excessive clutch slippage and rapid wear on clutch, linings, pressure plates, bearings and belts.

Clutch pedal free travel (Fig. 28) may be a guide to conditions inside the clutch. When the clutch is engaged, a retracting spring pulls the pedal back so that the release bearing fork does not contact the release bearing. Another spring pulls the release bearing carrier

Fig. 28 — Clutch Pedal Free Travel Adjustment

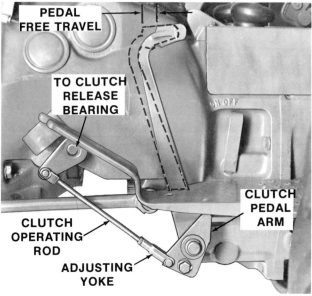

back so that the bearing does not contact the clutch plate levers. Thus, in normal driving, with the clutch engaged and the driver's foot removed from the clutch pedal, the release bearing is not turning — if pedal free travel is correct.

However, as clutch disk facings wear, the disk becomes thinner and plates move closer to the flywheel to engage the disk. This means the pressure plate must also move closer to the flywheel and release fingers move closer to the release bearing. After long operation, if not readjusted, these fingers will contact the release bearing and cause it to turn all the time.

But, additional wear on the clutch disk cannot be taken up by the pressure plate because the release fingers are contacting the release bearing. The clutch will then start slipping and the release bearing will wear out rapidly.

Adjust clutch pedal free travel any time the clutch has been repaired or whenever the clutch slips under load.

To determine pedal free travel:

● Slowly move the clutch pedal by hand (for better feel) from the fully engaged position until you start to feel resistance to pedal movement.

● Measure the distance the pedal moved (Fig. 3) and compare it to the free travel specified in the operator's manual or technical service manual.

● Adjust the length of the clutch operating rod (Fig. 28) to increase or decrease pedal free travel. The recommended free travel varies widely between vehicles, so always follow the manufacturer's instructions for each machine.

If proper pedal free travel cannot be obtained by adjusting external clutch linkage, the clutch must be adjusted or repaired internally. Also check the release yoke movement to verify that pedal free travel is actually the release bearing clearance, and not lost motion in worn linkage.

Other clutch linkage must be periodically adjusted too. For instance, as V-belts wear and stretch, it is usually necessary (depending on machine design) to adjust the linkage on belt tensioning clutches. Shortening the linkage (Fig. 29) permits full movement of the clutch control to ensure adequate tightening of the belt during operation to avoid slippage. When a new belt is installed, the linkage will usually need to be lengthened to compensate for the shorter initial belt length.

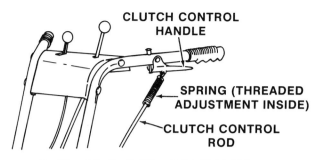

Fig. 29 — Adjusting Belt Tensioning Clutch Linkage

TROUBLESHOOTING CLUTCHES

Correct diagnosis and correction of problems of disk and plate clutches is generally more difficult than for other clutches, such as belt tensioning devices, which are simpler and more accessible. The following troubleshooting section is devoted to analysis of eight basic clutch troubles and their possible causes. Those eight troubles are:

● **Chattering** — especially in low or reverse gear.

● **Dragging** — failure to release promptly and fully, making it hard to shift gears.

● **Squeaks** — particularly when pedal is depressed.

● **Rattles** — especially at low speeds or standing.

● **Grabbing** — violent and sudden engagement.

● **Slipping** — failure to transmit full power.

● **Vibration** at high or low speeds or periodically.

● **Failure** to transmit power at all.

Note that different clutch problems may have the same cause. For example, oil or grease on clutch disk facings can cause slippage, chatter, drag or grab. Also, there could be oil on the linings, worn parts and improper linkage adjustment all causing trouble in the same clutch at the same time. So, correcting a single problem does not mean all the trouble is solved.

CHATTERING

● Oil or grease on clutch disk facings

● Glazed or worn facings

● Worn, loose, or spongy engine mountings

● Worn splines in clutch hub or on shaft

● Splined disk hub sticking on splined shaft

- Wear or looseness in universal joint, differential, or drive axles
- Cracked or scored pressure plate or flywheel face
- Warped clutch disk or pressure plate
- Pressure plate sticking on driving studs
- Sticking, binding, or unequally adjusted release levers
- Unequal length or strength of clutch springs
- Bent clutch shaft
- Misalignment of parts

DRAGGING

- Oil or grease in clutch
- Warped clutch disk, pressure plate, or clutch cover
- Splined disk hub sticking on splined shaft
- Sticking pilot bearing or bushing in flywheel
- Sticking release sleeve
- Broken disk facing
- Accumulation of dust in clutch
- Incorrect clutch or pedal adjustment
- Engine idling too fast
- Misalignment of parts

SQUEAKS

- Clutch release bearing or release sleeve needs lubrication
- Pilot bearing in flywheel needs lubrication
- Misalignment of parts

RATTLES

- Loose hub in clutch disk
- Worn pilot or release bearings
- Worn splines in hub or on shaft
- Worn driving pins in pressure plate
- Wear in transmission drive line
- Worn transmission bearings
- Bent clutch shaft
- Unequal adjustment of release levers
- Misalignment of parts

GRABBING

- Oil or grease on clutch disk facings
- Splined hub sticking on splined shaft
- Pressure plate sticking on driving studs
- Glazed or worn facings
- Sticking or binding release levers, clutch pedal, or linkage

SLIPPING

- Worn clutch disk facings
- Weak or broken springs
- Improper clutch or pedal adjustment
- Oil or grease on facings
- Warped disk or pressure plate
- Sticking release levers
- Pressure plate sticking or binding on studs
- Misalignment of parts

VIBRATIONS

- Bent clutch shaft
- Defective clutch disk
- Dust in clutch
- Improper assembly of clutch to flywheel
- Use of rigid disk instead of flexible
- Unmatched springs in pressure plate
- Misalignment of parts

FAILURE

- Clutch disk hub worn out
- Clutch friction facings torn off or worn out
- Broken springs
- Incorrect adjustment of pressure plate, clutch, or pedal
- Splined hub stuck on splined shaft

This chart can help you locate basic clutch problems. Additional specific problems may be described in the technical service manual, along with instructions for proper clutch repair and adjustment.

CHAPTER 4 REVIEW

1. Name four types of friction clutches.

2. (True or false) Belt tensioning clutches may be either normally engaged or normally disengaged until the operator moves the clutch pedal or lever.

3. (True or false) Jaw clutches are designed for smooth, gradual engagement.

4. Name two major types of disk and plate clutches.

5. (Insert two words) Releasing the clutch pedal permits the _____ _____ to force the pressure plate against the clutch disk.

6. What type of expanding shoe clutch is commonly found in compact equipment?

7. What feature do all friction clutches have in common?

8. What happens when the control for an electromagnetic clutch is "engaged"?

9. Name three common types of clutch controls, and state which is most common on compact equipment.

10. (True or false) Clutch pedal free travel is a standard dimension on all disk and plate clutches.

CHAPTER 5
MECHANICAL TRANSMISSIONS

POWER TRAINS
COMPACT EQUIPMENT

FUNDAMENTALS
OF SERVICE FOS

SKILLS AND KNOWLEDGE

This chapter contains basic information that will help you gain the necessary subject knowledge required of a service technician. With application of this knowledge and hands-on practice, you should learn the following:

• **Operation of variable speed belt drives.**

• **Service and maintenance of variable speed belt drives.**

• **Operation and maintenance of dual belt drives.**

• **Operation, service, and maintenance of friction-disk drives.**

• **Operation, service, and maintenance of friction cup and cone drives.**

• **Operation and maintenance of sliding gear transmissions.**

• **Operation of sliding gear transmission in conjunction with a drive clutch and driven clutch.**

• **Operation and maintenance of shifter key transmissions.**

• **Operation of shifter key transmission with wet multi-disk clutch/brake packs.**

• **Operation and maintenance of collar shift transmissions.**

• **Operation and maintenance of synchronizer shift transmissions.**

• **Types of shift controls used in gear transmissions.**

• **Troubleshooting belt-type transmissions.**

• **Troubleshooting gear-type transmissions.**

INTRODUCTION

As stated in Chapter 1, transmissions are devices for changing the speed or direction of machine operation. Variable speed and dual belt drives are widely used in lawn and garden equipment. Gear transmissions are often used in mowers and tractors of all sizes.

The types of mechanical transmissions most commonly found in compact equipment are listed below. The first three are belt or friction-type transmissions, and the last four are variations of gear-type transmissions.

- **Variable speed belt drives**
- **Dual belt drives**
- **Friction disk drives**
- **Sliding gear transmissions**
- **Shifter key transmissions**
- **Collar-shift transmissions**
- **Synchronizer shift transmissions**

Troubleshooting charts for these transmissions are included at the end of this chapter.

VARIABLE SPEED BELT DRIVES

The variator sheave (Fig. 1) on a variable speed belt drive moves like a pendulum between the engine drive sheave and the transmission sheave. As the movable center section of the variator (Fig. 2) slides all the way to the right, the belt on that side is squeezed up to the outside of the sheave. Simultaneously, the belt riding on the other side slides down toward the center of the sheave.

Moving the variator sheave toward the engine drive sheave (A, Fig. 1) slackens belt and disengages the drive. Shifting the variator to the right (away from the engine sheave) increases tension on the primary belt and engages the drive (B, Fig. 1).

As the variator moves further from the engine sheave, the primary belt is forced close to the center of the variator, thus increasing the rotating speed of the variator sheave. At the same time, the secondary belt moves further from the center of the variator, increasing its effective diameter, which provides an additional increase in speed to the transmission sheave. Maximum input speed to the transmission sheave is obtained when the variator sheave is farthest from the engine sheave. At this point, the primary belt is operating as close as it can get to the center of the variator, and the secondary belt is nearest the outer edge of the variator.

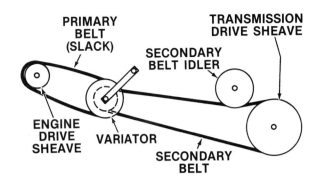

A. DISENGAGED POSITION

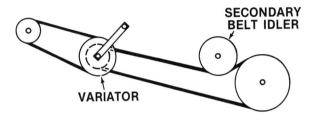

B. SLOW SPEED, HIGH TORQUE POSITION

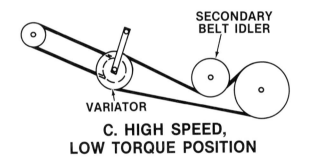

C. HIGH SPEED, LOW TORQUE POSITION

Fig. 1 — Variable Speed Belt Drive

Note that the engine must be running for belts to change position in the variator sheave. Trying to change the speed when the engine is stopped can damage belts or linkage.

Fig. 2 — Variator Sheave For Belt Drive

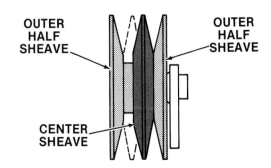

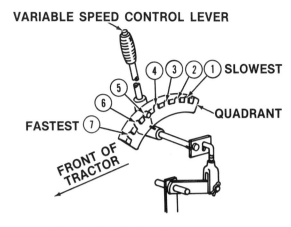

Fig. 3 — Variable Speed Control Lever

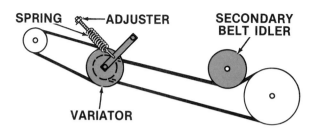

Fig. 4 — Torque Sensing Variable Speed Drive

If movement of the variator is lever controlled (Fig. 3), input speed to the transmission is varied in definite steps (position 1 is slowest, position 7 is fastest). However, the drive becomes load and torque sensitive if the variator arm has flexibility with respect to the variable speed control lever. As the load increases, added tension on the secondary belt (Fig. 4) pulls that belt deeper into the variator sheave, slowing speed and increasing torque to the transmission drive sheave. A spring attached to the variator arm is anchored to the machine by an eyebolt which permits adjusting sensitivity of the variator by increasing or decreasing spring tension. As the load decreases, the secondary belt moves outward in the variator, increasing transmission sheave drive speed and reducing the torque.

SERVICE AND MAINTENANCE OF VARIABLE SPEED BELT DRIVES

Normal belt wear and stretch will gradually reduce the effectiveness of variable speed drives. If the drive becomes inoperative at the lower end of the speed range (No. 1, Fig. 3), examine the belts for excessive wear and readjust the linkage.

Linkage adjustments usually involve several specific steps which must be followed in the proper sequence. So, always follow the instructions presented in the machine operator's manual or the technical service manual.

To adjust a typical variable speed belt drive (Fig. 5), follow these steps:

1. Disconnect the brake linkage at "C".

2. Place the variable speed control lever in the third notch from the front.

3. Disconnect the spark plug and turn the engine with the key starter several revolutions.

4. Measure distance "E" between bottom of the foot rest and the top of the clutch-brake pedal (0.50 inch (13 mm) on this tractor).

5. Adjust linkage by inserting a tapered punch or rod in the threaded rod "D" and turn "D" up or down until dimension "E" is correct.

6. Remove clevis pin "B" and while holding link "A" to the top of the slot, turn the threaded clevis up or down until the clevis pin can be inserted easily. Reinstall spring locking pin at "B".

7. Connect pin "C" temporarily.

8. Turn the engine several times with the key starter while moving the ground control lever to the rear (slowest position).

9. Depress the clutch-brake pedal as far as possible and measure distance between the pedal and top of the footrest, dimension "H" (insert Fig. 5). If distance is incorrect (0.75 inch (19 mm) for this tractor), turn brake rod into clevis "F" until the correct dimension is obtained at "H". Insert pin "C" and locking pin.

10. Turn nut "G" on parking brake rod either up or down until the clutch-brake pedal can be held in the lower position.

If, after adjusting linkage, the tractor still will not move when ground speed control lever is in the slow speed (rear) position, turn the threaded clevis up one or two turns on link "A". Install a new primary belt if necessary and repeat adjustments.

If the proper linkage adjustment fails to restore full-range speed control, replace the drive belts. Also check for adequate tension on the variator arm spring and be sure the variator is returning properly when the drive is accelerated.

To increase load sensitivity of the variator sheave, that is increase torque earlier under load, loosen the spring tension by lengthening the eyebolt (Fig. 6). For less load sensitivity, shorten the eyebolt to increase spring tension.

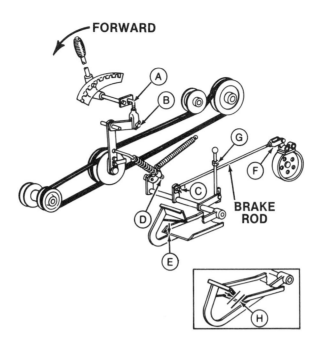

Fig. 5 — Adjustment Of Variable Speed Belt Drive

If the variator sheave fails to function smoothly, check sheave alignment and move the engine or transmission sheaves as necessary. Also check for dirt and debris in the bottom of sheave grooves. Replace bent, nicked, or badly worn sheaves.

Keep oil and grease off drive belts and remove any lubricants spilled on belts as soon as possible.

DUAL BELT DRIVES

Many walk-behind garden tillers have only one forward speed. However, these machines need a reverse drive to permit easier handling in tight corners and on row ends. This can be provided by a dual-belt drive.

Forward drive for such a machine is provided by a belt driven from a sheave on the engine crankshaft (Fig. 7). A belt tensioning clutch connects and disconnects power flow to the shaft which drives the tiller tines.

To reverse the drive, tension on the forward drive belt is released. Then the reverse drive control lever is engaged and tension applied to the second belt (driven in the opposite direction by a sheave on the engine camshaft). A dual sheave for the forward and reverse belts is mounted on the drive shaft to the tiller tines.

For safety, and because the reverse drive is used only intermittently, the reverse drive control must be held in place by the operator while the machine is backing up.

Fig. 6 — Adjusting Variator Spring

Fig. 7 — Dual-Belt, Forward-Reverse Drive

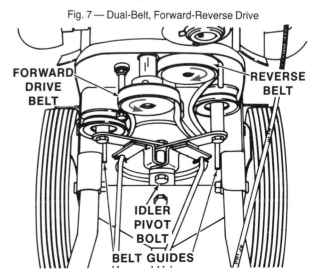

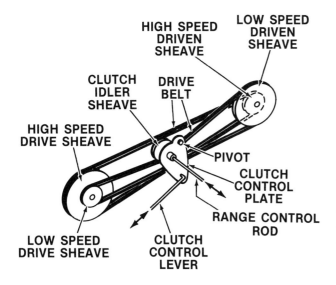

Fig. 8 — Parallel-Belt Drive

When the control is released, the belt becomes slack and neither drive is engaged.

PARALLEL BELT DRIVE

Two or more speeds can be provided by using two or more sheaves of different sizes in a V-belt drive (Fig. 8). For maximum versatility, a separate belt is installed for each set of sheaves. But, a single belt which must be shifted manually from one set of sheaves to the other is used in some parallel-belt drives.

To change speed, the idler clutch is engaged on a different drive belt. In Fig. 8, the clutch idler sheave can be aligned with either drive belt by sliding the range control rod back and forth through the clutch control plate. A spring-loaded ball inside the plate-sleeve engages detents in the sleeve to hold the idler in the desired range when belt tension is relaxed.

A parallel-belt drive may be used to provide two input speeds to a gear transmission.

SERVICE AND MAINTENANCE OF DUAL-BELT TRANSMISSIONS

Adequate tension is a must for satisfactory performance of any belt drive. If belts slip, adjust linkage as instructed in the operator's manual or technical service manual. If belts are badly stretched or damaged, or if linkage adjustment fails to stop slippage, replace the belts.

If one belt in a parallel-belt drive is used more or stretches faster than the other, it may become difficult to maintain proper control-idler adjustment when shifting the idler from one belt to the other. If both belts are initially the same length, it may be possible to interchange

them on the sheaves to equalize wear and improve performance.

Keep belts clean (free of grease and oil) and remove dirt from sheave grooves regularly.

FRICTION DISK DRIVES

Depending on the controls used, friction disk drives can provide infinite or stepped speed changes from zero to full throttle. For instance, with the friction disk (Fig. 9) rotating at a constant speed, the rotation speed of the traction wheel is fastest when it contacts the outer edge of the friction disk. The speed decreases if the traction wheel is moved nearer the center of the disk and, when the traction wheel is directly over the center of the disk, the wheel does not turn at all.

If the traction wheel is moved past the center of the friction disk, it begins to rotate again, but in the opposite direction. Wheel rotation speed increases again if the wheel is moved out toward the edge of the disk.

If reverse is not required, lateral movement of the traction drive wheel can be limited to one side of the friction disk.

In most friction disk drives, linkage is provided to lift the traction wheel off the disk to disengage the power flow. If the friction disk has an abrasive surface to increase traction, the drive wheel must be lifted and the disk stopped before changing speed. If not raised, lateral movement of the traction drive wheel across the friction disk will scuff or tear rubber from the traction drive wheel.

SERVICE AND MAINTENANCE OF FRICTION-DISK DRIVES

Bearings supporting the friction disk and traction wheel must be lubricated as specified in the operator's manual unless they are equipped with sealed bearings. However excess lubricant can make the disk and wheel slip.

Fig. 9 — Friction-Disk Drive

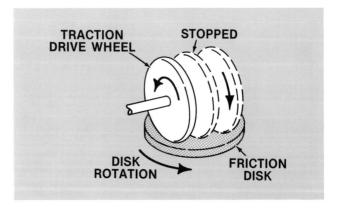

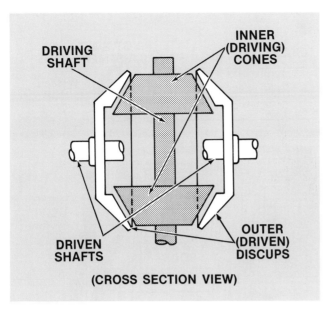

Fig. 10 — Friction Cup And Cone Drive In Neutral

If either the friction disk or traction drive wheel wear so much they can't maintain a positive drive, they must be replaced. Refer to the technical service manual for repair instructions and additional service information.

FRICTION CUP AND CONE DRIVES

A unique friction cup and cone drive is used to move some riding mowers. The basic drive consists of two rotating cones mounted on a fixed vertical shaft and two movable friction cups (Fig. 10). The cones are driven by a V-belt from the engine. The cups are mounted on individual shafts which transmit power to each drive wheel.

A separate steering lever controls the movement of each cup, and in turn the speed and power flow to each drive wheel. So, when both levers are in "neutral" (A, Fig. 11), neither cup contacts a drive cone and no power flows to the wheels.

If the cup on each side contacts the driving cone at the smallest diameter of the cone (both levers moved slightly forward), the cup (and the machine) will be driven at the slowest speed (B, Fig. 11). Tilting the cups in relation to the cones so that contact is made at the largest diameter of the cones (both levers full forward) produces the fastest speed (C, Fig. 11). If only one cup contacts a driving cone (D, Fig. 11 — left lever forward, right lever in neutral), the other drive cone (and drive wheel) will stop, and the machine will make a 90 degree right turn with the right wheel stationary.

Moving one lever forward and pulling the other one back an equal amount will cause one cup to drive forward and the other in reverse (not shown in Fig. 11). The machine will then spin in its own length. However, for safety and to avoid stressing the machine, speed should be reduced before making such a maneuver.

Minor steering changes are accomplished by shifting one control lever slightly forward or back in relation to the other one. This does not necessarily shift one side into reverse; it only changes the speed at which each wheel travels.

Pulling both levers slightly rearward an equal amount provides a slow reverse speed (E, Fig. 11). This is the opposite cup cone contact situation found in B, Fig. 11. So, the cup and cone drive combines the functions of a clutch (to start and stop power flow), a transmission (to change speed and direction of operation), and a differential (to permit drive wheels to change speed when turning).

Fig. 11 — Operation Of Friction Cup And Cone Drive

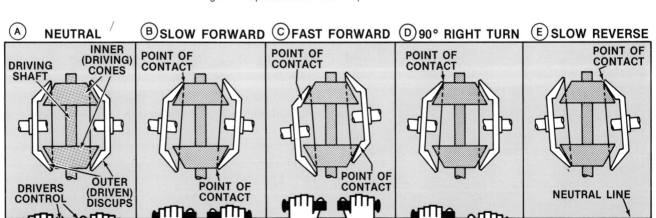

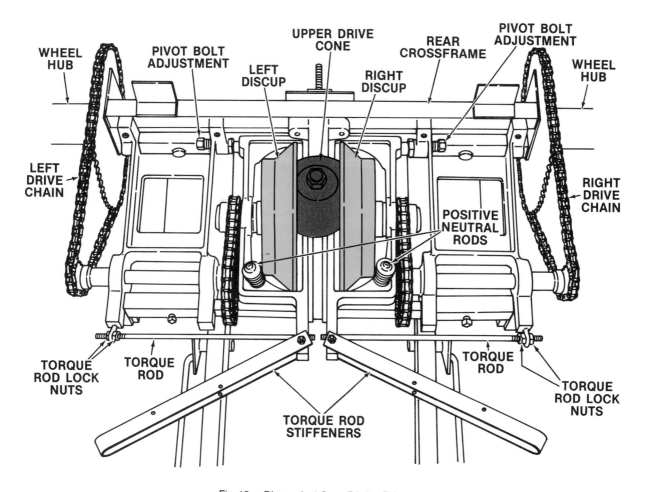

Fig. 12 — Discup-And-Cone Friction Drive

SERVICE AND MAINTENANCE OF FRICTION CUP DRIVES

The V-belt drive and chain drives to the wheels must be properly tensioned for efficient operation. Replace the belt or chains if they become so loose that the slack cannot be removed by adjusting the belt or chain tighteners.

For proper operation, the cups must be set parallel to each other and the center frame and perpendicular to the rear crossframe (Fig. 12).

The clearance between the cups and the upper cone must be equal for smooth operation. When checking cup clearance (Fig. 13), move each cup upward until it contacts the bottom cone. The recommended clearance is 0.030 to 0.050 inch (0.762 to 1.27 mm), but **making sure each cup has equal clearance is more important than the amount of clearance.**

To change the cup-to-cone relationship, adjust torque rod lock nuts (Fig. 12) to shift the cups closer to or away from the cones. To keep cups parallel when torque rod length is adjusted:

● *Change the pivot bolt adjustment too (Fig. 12).*

● *Make one turn of the pivot bolt nut for each two turns made on the torque rod nuts.*

● *After adjusting the cups, tighten pivot bolt and torque rod lock nuts.*

Make the fine adjustment of cup clearance by adjusting the positive neutral rods (Fig. 12) until the spacing is the same on both sides between each cup and the top cone (Fig. 13). Springs on the positive neutral rods must be depressed enough to return control levers to neutral when they are released.

Control levers should be parallel when in the neutral position and in the full speed forward position with the machine running straight. Telescoping control rod assemblies can be lengthened or shortened as necessary so that levers are parallel. (A difference of one inch (25 mm) between right-hand and left-hand levers when the machine is running straight forward at full speed is acceptable.)

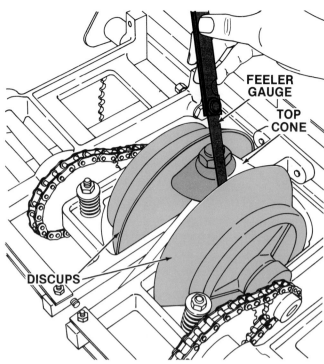

Fig. 13 — Check Cup-To-Cone Clearance

When adjusting control rods, reduce speed of the faster wheel as necessary to match the top speed available from the slower speed when traveling straight ahead. Large variations in speed of the two wheels indicates improper setup or damaged or worn parts. If the machine goes to the left, the right wheel is traveling faster than the left wheel. So, lengthen the left control rod and shorten the right rod.

Adjusting the torque rod stiffeners will compensate for minor differences if control levers are not parallel when the drive is in neutral.

If the above adjustments fail to provide parallel lever positions at full speed and in neutral, look for defective, damaged, or worn parts and replace them if necessary.

For major drive repairs, remove the drive unit from the machine. Special tools and equipment will be necessary. They are available from machine manufacturers.

GEAR-TYPE TRANSMISSIONS

The **operation** of the gear-type transmissions listed on page 5-3 will be covered in the following sections.

Shift controls for these types of transmissions will also be discussed and will be followed by **service and maintenance** procedures.

SLIDING GEAR TRANSMISSIONS

Sliding gear transmissions are common in farm and industrial machines as well as compact equipment. They consist primarily of spur gears and shafts, are simple to operate and repair, and can provide a variety of speeds.

There are two basic designs:

● **Input and output shafts parallel**

● **Input and output shafts in line**

Input and Output Shafts Parallel

Power flows from the input shaft, through various combinations of gears, to the output shaft which transmits power to the differential or other outlet. There is normally a third shaft (counter or idler shaft) which reverses power flow to the output shaft. All three shafts are parallel. A typical arrangement (Fig. 14) provides three forward speeds and one reverse speed.

Gears D and E (Fig. 14) are splined to the output shaft and slide from one position to another to mate with gears on the input shaft. All other gears are fixed to their respective shafts except gear F, which turns freely on the output shaft.

The dotted lines show the gear positions when they are shifted.

Engaging gear D with gear A provides first or low gear (a small gear driving a larger gear). When gear D is disengaged from A, and gear E is engaged with gear B, the transmission is in second gear (gears of approximately equal size provide an in-between speed).

Engaging gear E with gear F causes power flow from the input shaft through gear C to gear F and the output shaft. This provides third or high gear (a large gear driving a smaller gear increases output speed).

Gears G and H are fixed on the reverse idler shaft and gear H is in constant mesh with gear A. Therefore, gear G rotates whenever the input shaft is turning. Engaging gear D with G reverses the rotation of the driven shaft to the differential.

If neither gears D or E are engaged with another gear, the transmission is in neutral. The shift linkage in a sliding gear transmission is designed to always disengage any pair of mating gears **before** another pair can be engaged. This prevents gears working against each other in trying to drive the output shaft at two speeds at the same time.

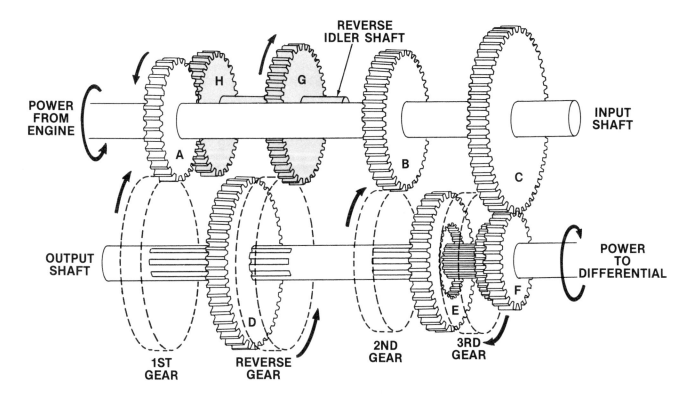

Fig. 14 — Sliding Gear Transmission With Parallel Shafts

Fig. 15 — Drive Clutch Operation

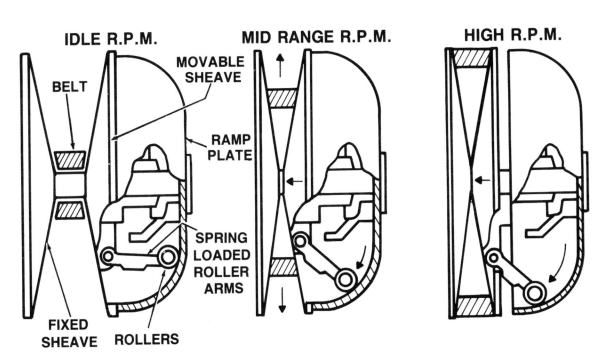

A drive clutch and driven clutch may be used in conjunction with a sliding gear transmission to provide engine speed and, load and torque sensitive operation. Let's look at how the clutches operate.

DRIVE CLUTCH

The fixed sheave face (Fig. 15) is keyed to the engine crankshaft. Attached to the fixed sheave at the other end of the clutch hub is the ramp plate. The moveable sheave and the spring arm assembly ride on the clutch hub between the fixed sheave and the ramp plate.

The drive clutch is rpm sensitive and operates using the principal of centrifugal force. Increasing engine rpm causes the spring-loaded roller arms and rollers to swing out due to centrifugal force. The roller arms are attached to the movable sheave. As these roller arms are forced out they contact the ramp plate. The roller arms follow the contour of the ramp plate and force the movable sheave toward the fixed sheave. This action causes the belt to ride up between the two sheave surfaces in effect changing the working diameter of the drive clutch.

At idle rpm, the clutch sheaves are farthest apart and the drive belt is located at the bottom of the sheave halves. The sheave halves do not make contact with the drive belt (Fig. 16) so there is no movement.

At midrange rpm, the roller arm assemblies have moved further down the ramp plate causing the movable sheave to move closer to the fixed sheave. This forces the belt further toward the top of the clutch assembly.

At high rpm, the roller arms have moved against the ramp plate the full extent of their travel forcing the sheave halves together. This results in the drive belt being moved to the largest diameter of the drive clutch.

DRIVEN CLUTCH

The shift pattern is controlled in order for the engine to operate efficiently throughout the entire speed range. The shift ratio limits engine rpm. If engine speed is too high, it will be operating over the engine power peak. If engine rpm is too low, the power will be inadequate for the load.

At the recommended rpm, the drive clutch engages the belt. The belt applies torque to the drive clutch (Fig. 17). At engagement rpm, the ratio between the drive sheave and the driven sheave is the highest.

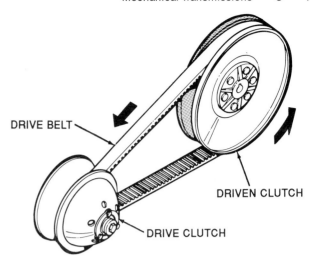

Fig. 16 — Drive Belt Connects Drive Clutch And Driven Clutch

As the belt engages, torque from the belt is felt on the movable sheave and torque from the resistance of the track is felt on the stationary sheave.

As the engine rpm increases, the action of the drive clutch forces the belt to its outside diameter. If only a light load is encountered, sheave rpm overcomes the spring in the driven clutch and the movable sheave rides down the ramp opening the driven sheave halves. This in effect, reduces the working diameter of the driven clutch and at the same time increases the working diameter of the drive clutch. The drive ratio from engagement rpm to wide open throttle rpm changes from high to low. This action will continue until the clutches have reached the lowest drive ratio. This action is referred to as *upshift*.

Whenever a load is encountered, the clutch system reacts and automatically changes the drive ratio to a higher ratio. This allows the engine to operate at optimum efficiency at the peak of its power curve. Load is increased by anything that adds resistance to the wheels. Examples include going from flat to hilly terrain, or from a hard surface to soft terrain.

As soon as increased load is encountered, the stationary sheave of the driven clutch starts to resist forward movement. At the same time, the movable sheave continues to have forward torque applied by the belt. The movable sheave of the driven clutch moves forward, assisted by the spring. As it moves forward, the ramp causes it to move toward the stationary sheave forcing the belt toward the outside diameter of the sheave halves, increasing the drive

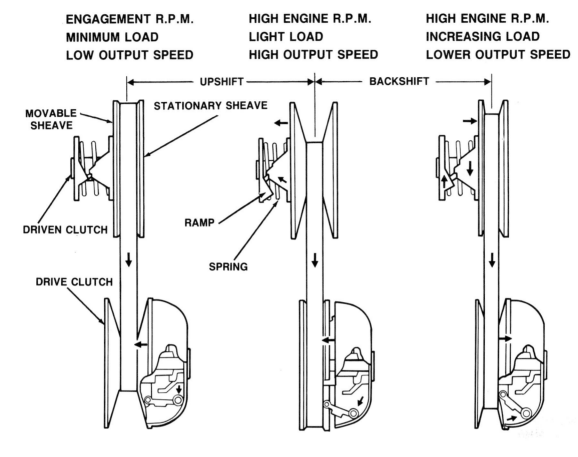

ENGAGEMENT R.P.M.   HIGH ENGINE R.P.M.   HIGH ENGINE R.P.M.
MINIMUM LOAD   LIGHT LOAD   INCREASING LOAD
LOW OUTPUT SPEED   HIGH OUTPUT SPEED   LOWER OUTPUT SPEED

UPSHIFT   BACKSHIFT

MOVABLE SHEAVE   STATIONARY SHEAVE

DRIVEN CLUTCH   RAMP

SPRING

DRIVE CLUTCH

Fig. 17 — Drive Clutch And Driven Clutch Operation

ratio and belt tension is maintained, while vehicle speed is reduced. This action is referred to as *backshift.* The clutches will continue to backshift until the load stabilizes.

As the load is reduced, the clutches will again upshift to a ratio dependent on engine rpm. Upshift and backshift occur continually throughout operation dependent on load, or engine rpm changes.

Input and Output Shafts In Line

In this transmission (Fig. 18), input and output shafts lie in a straight line but are not connected. A third shaft, or countershaft, transmits power between them.

Gears D, E, F, and G are fixed on the countershaft. To obtain low or first gear (Fig. 18), power flows from gear A on the input shaft to gear D on the countershaft. This drives gear F which is engaged with gear C which is splined to the output shaft. This causes the output shaft to rotate and transmit power to the differential.

Second gear is obtained by engaging gear B on the output shaft with gear E on the countershaft. However, there is a direct flow of power for high gear when gears A and B are connected thus locking together the input and output shafts.

Numerous variations of these sliding gear transmissions are used in different machines. They may provide as many as 10 forward speeds and two reverse speeds, depending on machine and transmission design.

SHIFTER KEY TRANSMISSIONS

The transmission input sheave (Fig. 19) is keyed to the input shaft. Splined to the shaft is the input shaft gear that drives the bevel input gear that is splined to the reduction shaft. Also splined to the shaft is the forward reduction gears, and the reverse reduction sprocket. Therefore, whenever the input shaft is turning, so are the reduction gears and sprocket. The shifter gears are in constant mesh with the reduction gears and the reverse shifter shaft sprocket is direct driven by the chain. These gears and sprocket float freely on the shifter shaft.

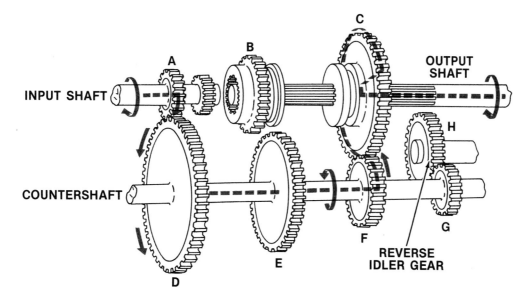

INPUT SHAFT

OUTPUT SHAFT

COUNTERSHAFT

A B C

D E F G H

REVERSE IDLER GEAR

Fig. 18 — Sliding Gear Transmission With Shafts In Line (First Gear Shown)

Fig. 19 — Shifter Key Transmission In Neutral

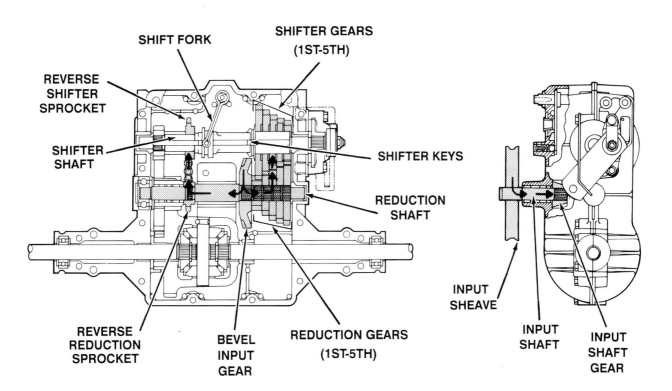

SHIFT FORK

SHIFTER GEARS (1ST-5TH)

REVERSE SHIFTER SPROCKET

SHIFTER SHAFT

SHIFTER KEYS

REDUCTION SHAFT

REVERSE REDUCTION SPROCKET

BEVEL INPUT GEAR

REDUCTION GEARS (1ST-5TH)

INPUT SHEAVE

INPUT SHAFT

INPUT SHAFT GEAR

MOVING COMPONENTS

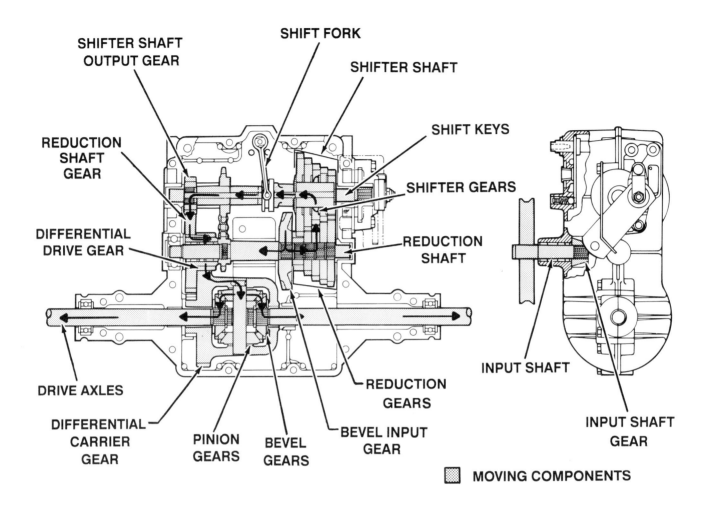

Fig. 20 — Shifter Key Transmission In Forward-3rd Gear

The shifter shaft has four machined keyways that hold the shifter keys. These keys are actuated by the shift fork which engage the keys into pockets of a selected gear. When engaged in the pocket of a gear, power is transferred from the shifter gears to the shifter shaft.

Neutral (Fig. 19)

In neutral, the shift fork places the keys outside the gears. Therefore, no power is transmitted to the shifter shaft and neutral is attained.

Forward (Fig. 20)

By moving the shift fork the keys engage in pockets of a selected gear. Power from the shifter gear is transferred to the shifter shaft through the keys. The shifter shaft output gear transfers power to the reduction shaft gear which is splined to the differential drive gear that mates with the differential carrier gear.

The differential carrier transfers power to the drive axles through the differential pinion gears that mate with the axle bevel gears.

Reverse (Fig. 21)

When the transmission is placed in reverse (Fig. 21), the shift fork moves the keys into pockets in the shifter shaft reverse sprocket. This allows power to be transferred from the reduction shaft sprocket through the chain to the shifter shaft sprocket. The sprocket transfers power through the keys to the shifter shaft which now turns in the opposite direction as it did in any forward gear.

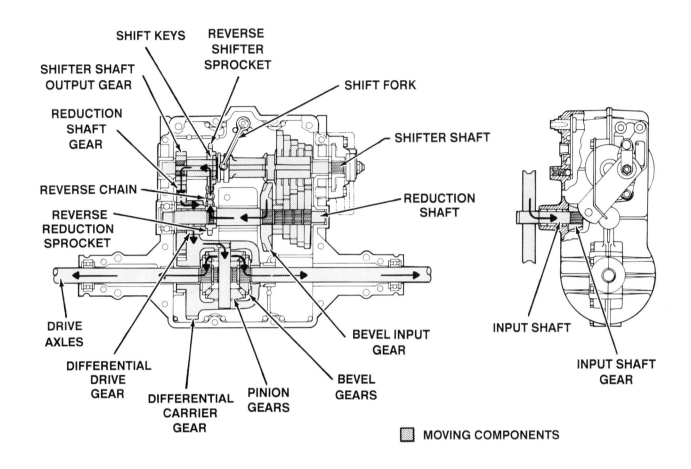

Fig. 21 — Shifter Key Transmission In Reverse Gear

The shifter shaft output gear transfers power to the reduction shaft gear which is splined to the differential drive gear that mates with the differential ring gear.

The differential carrier transfers power to the drive axles through the differential pinion gears that mate with the axle bevel gears.

The transmission may be equipped with wet, multi-disk clutch/brake packs to propel and brake the drive wheels. Let's look at how that transmission operate.

Shifter Key Transmission With Wet Multi-Disk Clutch/Brake Packs

Power is transmitted from the belt drive to the input shaft of the transmission. The input shaft transmits power from the input pinion gear to the bevel gear (Fig. 22) that is splined to the reduction shaft.

Also splined to the reduction shaft are the reduction gears for first through fifth, the neutral collar and the reverse gear. Therefore, whenever the input shaft is turning so are all of the gears on the reduction shaft.

The mating shifter gears on the shifter hub are in constant mesh, but float freely. The shifter hub incorporates four grooves containing spring-loaded shift keys that engage into pockets in the selected gear. The selected gear then transfers the power through the shift key to the shift hub.

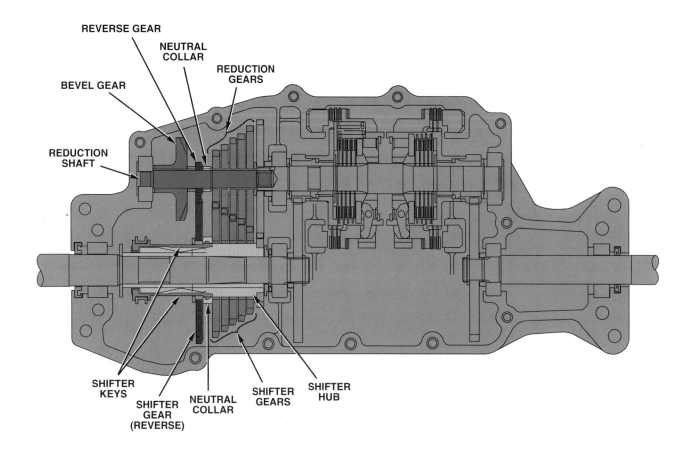

Fig. 22 — Shifter Key Transmission With Wet Multi-Disk Clutch/Brake Packs In Neutral

NEUTRAL (Fig. 22)

When the transmission is in neutral, the shift keys are not engaged in any gear, but are positioned under the neutral collar. Therefore, no power is transferred to the shift keys and beyond.

FORWARD (Fig. 23)

When the gear selector is shifted into a forward gear, the shifter keys engage in the pockets of the selected gear. The gear set now transfers power through the keys into the shifter hub.

A two-sided drive cog at the end of the shifter hub drives the gear splined to the end of the clutch shaft. When the clutches are engaged, the clutch output gears drive the axle bull gears that power the wheel axles.

REVERSE (Fig. 24)

To achieve reverse, the shift collar engages the shift keys in the reverse gear of the shifter hub. The reduction shaft reverse gear meshes with an idler gear. The idler gear then meshes with the shifter

hub reverse gear. This extra gear is needed to change the rotation of the shifter hub which provides reverse operation.

Because the shift keys are engaged in the shifter hub reverse gear, power is now transferred through this gear.

A two-sided drive cog at the end of the shifter hub drives the shifter hub output gear. This drives the clutch shaft input gear that is splined to the end of the clutch shaft. When the clutches are engaged, the clutch output gears drive the axle bull gears that power the wheel axles.

CLUTCH/BRAKE PACKS (Fig. 25)

The clutch/brake packs provide three functions:

- **When engaged, power is transferred from the reduction gears to the wheel axle gears.**

- **When disengaged, power is interrupted between the reduction gears and the wheel axles.**

- **Apply braking action to the wheel axles.**

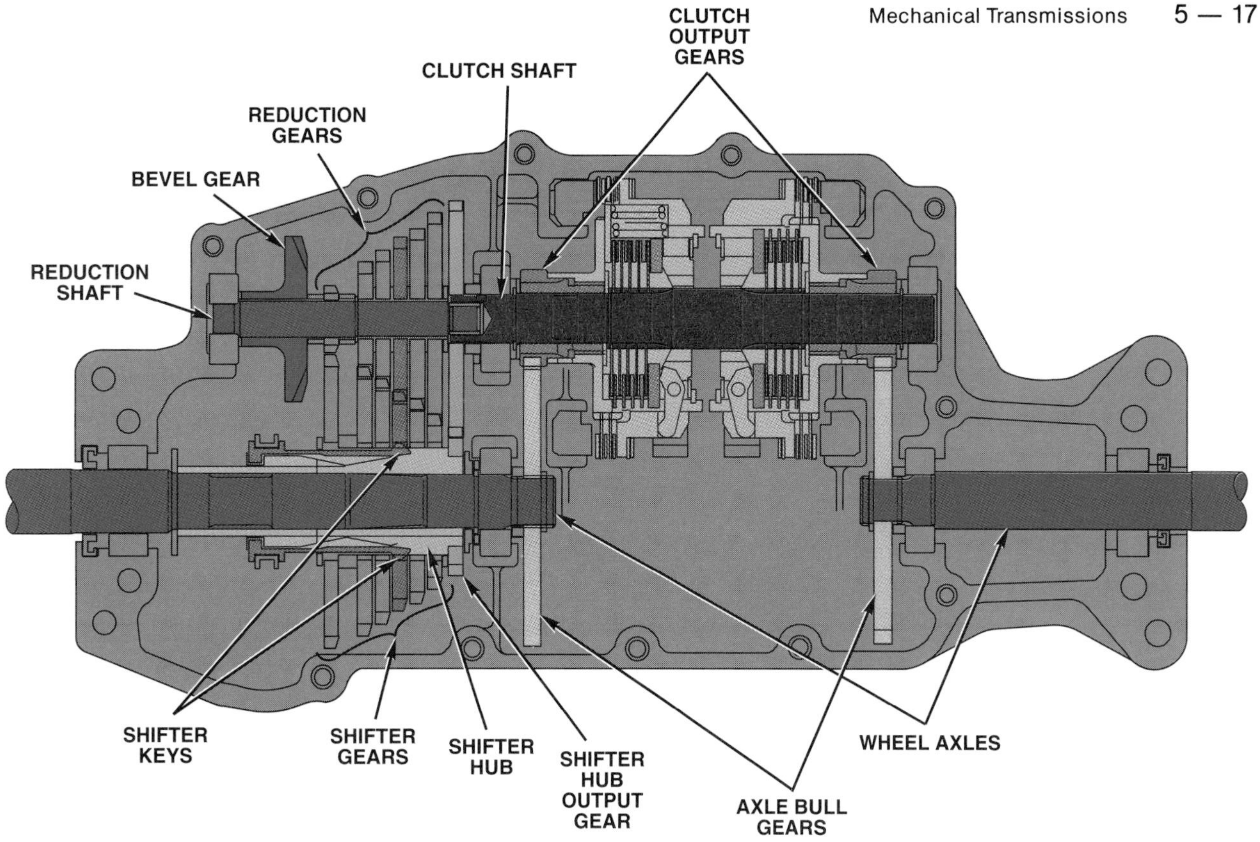

CLUTCH
OUTPUT
GEARS

CLUTCH SHAFT

REDUCTION
GEARS

BEVEL GEAR

REDUCTION
SHAFT

SHIFTER
KEYS

SHIFTER
GEARS

SHIFTER
HUB

SHIFTER
HUB
OUTPUT
GEAR

AXLE BULL
GEARS

WHEEL AXLES

Fig. 23 — Shifter Key Transmission With Wet Multi-Disk Clutch/ Brake Packs In Forward Gear

When engaged, the clutch packs transfer power. When operating in any gear, the clutch shaft is rotating. Splined to the clutch shaft are the drive discs. Sandwiched between the rotating drive discs are the separator plates. Tangs on the separator plates connect them to the clutch housing.

The clutch springs apply pressure to the cam plate housing. The pressure pivots the cam fingers down against the backing plate and forces the separator plates and drive discs together. Since the drive discs are splined to the clutch shaft, and the separator plates are tanged to the clutch housing, all components rotate as one assembly when engaged.

Splined to the end of the clutch housing is the clutch output gear that, with the clutch engaged, drives the axle bull gear.

Disengagement takes place when the clutch fork rotates, and applies force to the cam plate housing. The force from the clutch fork overcomes the clutch spring tension and the cam plate housing compresses the springs.

When the springs are compressed, the cam fingers no longer apply pressure to the backing plate. The drive discs continue rotating and the separator plates that drive the clutch housing are allowed to stop.

As the clutch fork continues movement beyond disengagement, braking action takes place.

Circular brake plates are attached to the clutch housing through external tangs on the plates. Therefore, whenever the clutch housing rotates the brake plates rotate also. Sandwiched between these brake plates are the brake disks. The disks remain stationary because of tangs that are located in pockets of the transmission case.

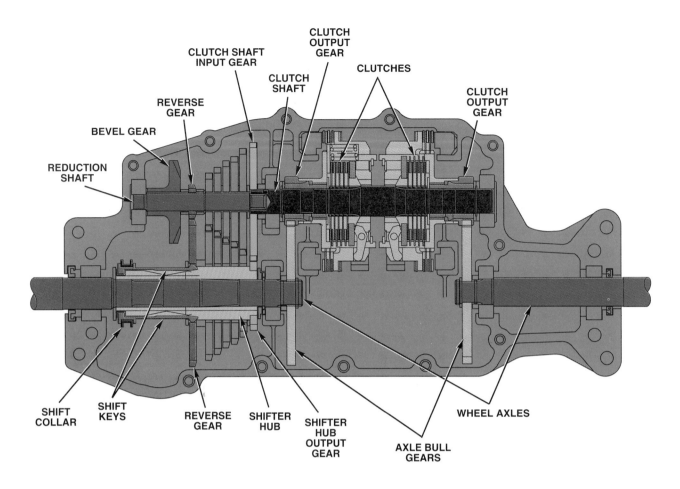

Fig. 24 — Shifter Key Transmission With Wet Multi-Disk Clutch/Brake Packs In Reverse Gear

As the clutch fork applies pressure, the disks and plates are squeezed between the cam plate housing and the brake adjuster braking the clutch housing and output gear.

COLLAR SHIFT TRANSMISSIONS

Collar shift transmissions have parallel shafts with the gears in constant mesh. In neutral, the gears are free-running. But when shifted they are locked to their shafts by sliding collars. Because the gears are constantly in mesh, quieter-running helical gears can be used in these transmissions. These gears have teeth cut at an angle (Fig. 26 driven gear).

When a gear is engaged, it is secured to its shaft by a sliding collar or coupling. When the collar is slid over, the gear is disengaged and released to turn freely.

A typical driven gear (Fig. 26) has short splines machined on one side of the gear. The shifter collar has matching internal splines and is splined to a shifter gear, which is splined to the shaft. Or, the collar may be splined to the shaft without a shifter gear.

Power is transmitted when the collar is shifted so its internal splines engage with the external splines on the shifter gear and driven gear. (Splines on the gears and collar have rounded ends for easier shifting.)

Let's see how the whole collar shift transmission works. Power enters the transmission (Fig. 27) from the engine. For first gear, gear 1 is locked to the shaft by collar A. Power then flows, as shown by arrows, from gear 1 through gear 3 to gear 4, and from there to gear 2 which is locked to its shaft by collar B.

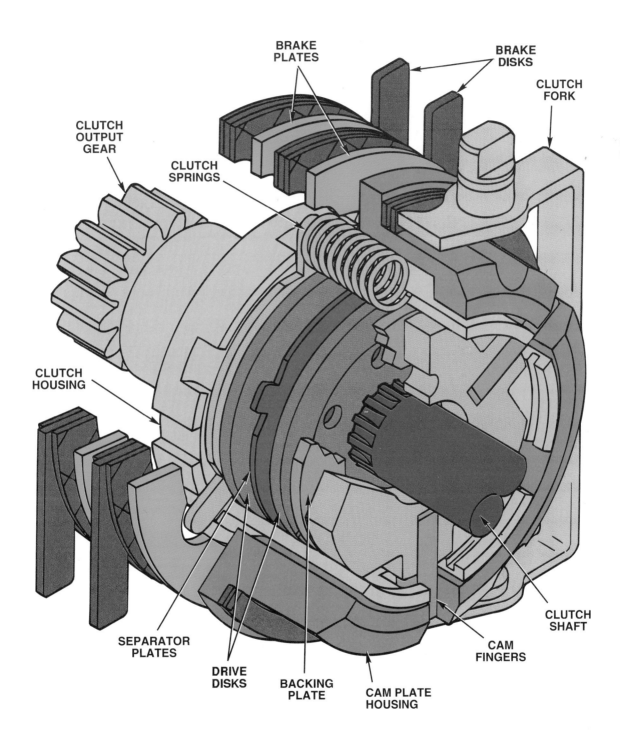

Fig. 25 — Wet Multi-Disk Clutch/Brake Pack In Shifter Key Transmission

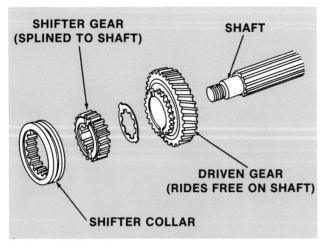

Fig. 26 — Typical Shifter Collar And Mating Gear

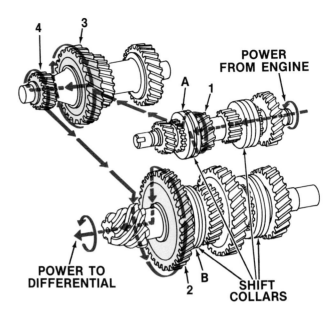

Fig. 27 — Collar Shift Transmission — Power Flow In First Gear

Fig. 28 — Power Flow Of Synchronizer Shift Transmission

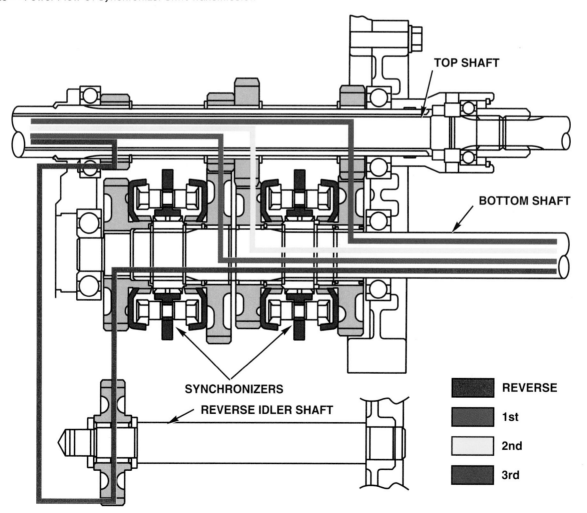

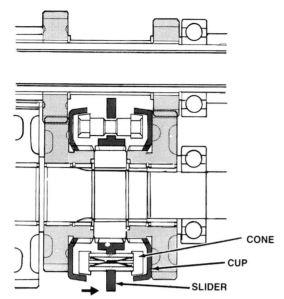

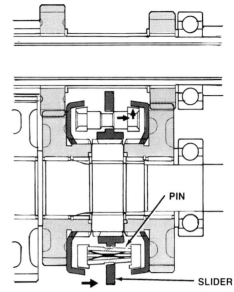

Fig. 29 — Beginning Of Synchronizer Shift

Fig. 30 — Midway Through Synchronizer Shift

(The shafts in Fig. 27 are separated to illustrate power flow. In a real transmission, gears 1 and 3 would be in constant mesh, as would gears 4 and 2.)

All eight speeds in the transmission shown in Fig. 27 are obtained by using the collars to secure the different combinations of gears to the shafts.

SYNCHRONIZER SHIFT TRANSMISSIONS

Synchronizer shift transmissions have parallel shafts with the gears in constant mesh. The gears on the top shaft (Fig. 28) are splined to the shaft. The gears on the lower shaft float free until the synchronizer is moved toward it. This engages the gear to the lower shaft through the synchronizer. Power will then flow from the top shaft through the gear engaged by the synchronizer and out the lower shaft. The synchronizer allows for shift "on-the-go" as it will equalize the speed of the mating parts before they engage.

As the slider in the synchronizer is moved to the right, the cone (Fig. 29) contacts the cup on the gear being shifted into. This will slow down or speed up the gear to match the speed of the synchronizer.

Continued force on the slider (Fig. 30) begins to move the slider over the pins in the direction of the gear. The pins now put spring tension on the slider to assist in forcing the slider and inner collar toward the gear.

After the gear and the synchronizer have matched speeds and the pins have put force on the slider, the inner collar (Fig. 31) of the synchronizer can easily and smoothly engage with the splines on the gear. When this happens, the slider and the inner collar move completely toward the gear. Power flow is now from the gear through the inner collar to the lower transmission shaft.

Fig. 31 — End Of Synchronizer Shift

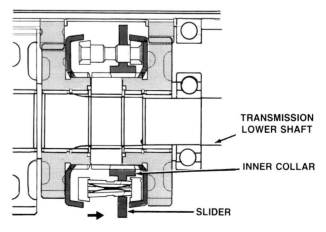

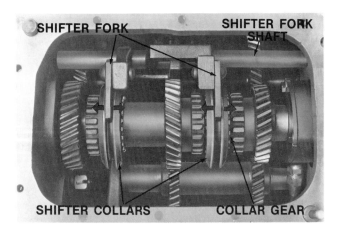

Fig. 32 — Shifter Forks In Collar Shift Transmission

SHIFT CONTROLS IN GEAR-TYPE TRANSMISSIONS

A shifter fork (Fig. 32) which fits into a groove in the collar or gear is used to change speeds in sliding gear, collar shift, and synchromesh transmissions. The shifter fork is connected to and moved manually by the gear-shift lever. Two common mechanisms are used to connect the gearshift lever with the shifter fork. These are:

- **Direct shifters.**

- **Cam shifters.**

Direct Shifters

In the direct shifter, the gearshift lever passes through the transmission cover from the operator's station to the shifter forks inside. A ball on the gearshift lever fits into a socket in the transmission cover (Fig. 33). This permits the lever to swivel into various shift positions. As the lever is moved, the finger on the lower end selects and moves one of the rails, its forks, and a gear or collar.

For example, in a four-speed transmission (Fig. 34), as the gearshift lever moves from neutral to the left, the lower end of the lever moves into the slot in the low-and-second speed rail. This action selects the rail, forks, and gears to be moved. Then, as the lever is pushed forward into low speed, the rail is pushed to the rear, engaging low gear.

When you look inside a direct-shift transmission, keep in mind that shift motions inside are opposite to movement of the lever. Moving the top of the gearshift lever forward moves the lower end to the rear, and vice versa. Moving the lever to the left shifts the lower end to the right, etc.

Fig. 33 — Direct-Type Transmission Shifting Mechanism

A popular method of holding shift rails in position and preventing other rails from moving uses spring-loaded detent balls under the rails. As the rails are moved, balls snap into detents to lock gears in mesh or hold other rails when different gears are used.

Refer to the technical service manual for details of the locking arrangement used in a particular transmission.

Cam-Type Shifters

A cam-type shifter (Fig. 35) may be used where it is inconvenient or impossible to operate a direct-type shifter — for example inside a tractor cab where it is undesirable to have the gearshift lever protruding through the floor.

In a two-cam arrangement (Fig. 35), one cam shifts the range gearing (high, low, reverse) and one is for the speed gearing (first, second, etc.). Moving the shift lever from the operator's station moves the cam about its pivot point. As the cam moves, shifter rollers follow grooves in the cam and move the shifter forks.

The irregular shape of the grooves in the cam moves the forks to engage the desired gear or range. The cam shifter also uses detents in the cam itself to lock the transmission in the desired gear.

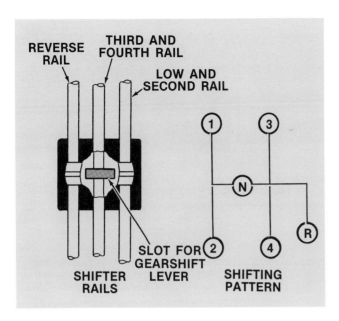

Fig. 34 — Shift Pattern Of A Four-Speed Transmission

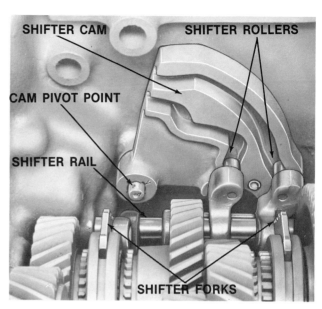

Fig. 35 — Shifter Cam And Rail Assembly

SERVICE AND MAINTENANCE OF GEAR-TYPE TRANSMISSIONS

Gear-type transmissions require little maintenance other than periodic oil changes.

But, when a transmission is disassembled for replacement of bearings or gears, some parts may require preloading or endplay adjustment as explained in Chapter 2. Specific procedures and measurements are supplied by the manufacturer in the technical service manual.

So, when repairing a transmission, always examine the entire gear train to locate worn or faulty parts and repair or replace them at the time. You may thus prevent a later breakdown and avoid the need to disassemble the transmission again soon.

TRANSMISSION REPAIR

If the transmission must be disassembled for repairs, drain the transmission oil. After repairs are made, refill the transmission with new oil of the type recommended by the manufacturer. Reusing old oil could contaminate the transmission with metal particles worn from parts and result in rapid wear and repeated failure.

Note: Special filtering systems are available for cleaning impurities from used transmission oil. However, this should only be done when the oil has not been used long enough to deplete the additives in the oil. If such equipment is used, follow carefully the manufacturer's instructions and avoid reusing oil from which additives have been lost.

For certain very simple repairs it may only be necessary to remove the gearshift retainer cap from the top of the transmission case (Fig. 36). However, most repairs will also require removal of the main cover and possibly other parts such as the differential lock cover and perhaps the rockshaft housing. Follow the steps outlined in the technical service manual and avoid removing any unnecessary parts. Some parts may be damaged as they are removed or replaced and thereby unnecessarily increase repair time and cost if removal is not required.

Major transmission repairs usually require separating the transmission case from the clutch housing (Fig. 37). Be sure to disconnect oil lines, wiring, and other parts between the two sections before separating the machine components.

If the transmission case is to be completely disassembled, or if one or more of the final drive (axle) housings is to be removed, use a disassembly stand (Fig. 38). This will provide stable support for the transmission while it is being repaired and reduce the possibility of personal injury or transmission damage if the unit should fall.

After removing the transmission cover, remove the detent springs (Fig. 39). Then use a small magnetic pickup tool to remove detent balls from detent bores. Remove the seal cap retainer plate and the reverse shift stop plate. Drive spring pins from shifter forks (bend a short piece of light wire to catch pins as they are driven out to keep them from

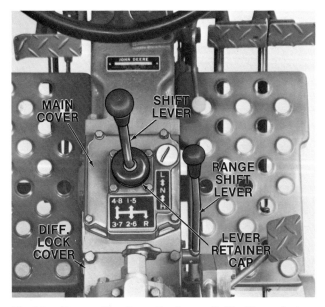

Fig. 36 — Remove The Cover For Access To The Transmission

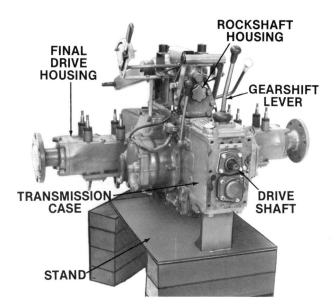

Fig. 38 — Support The Transmission On A Stand For Disassembly

dropping into the case) and then remove one shifter shaft at a time. To simplify the job, remove the shafts in the order suggested in the technical service manual. Be careful to avoid damaging shaft surfaces as the shafts are removed or reinstalled.

Use special tools and step-by-step procedure to remove transmission drive shaft and gears (Fig. 40). The spacers prevent gears from binding against each other as the shaft is removed and reduce danger of damaging parts. Be sure to place shift levers in the proper position during removal of various parts because this prevents other parts from interfering with those being removed.

Fig. 39 — Remove Shifter Shafts And Shifter Forks

Fig. 37 — Separate The Machine For Access To The Transmission

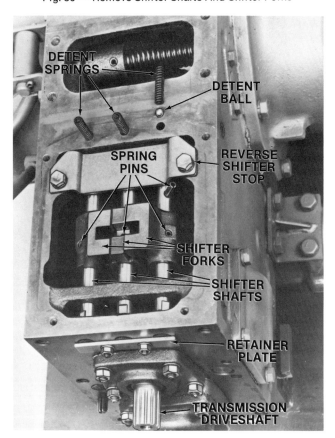

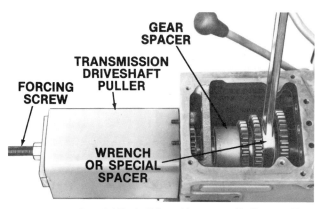

Fig. 40 — Use Special Tools Where Necessary To Remove Parts

IMPORTANT: If the differential drive shaft (Fig. 41) is replaced, the differential ring gear must also be replaced because these parts are supplied as a matched set.

After removing all necessary parts, clean the inside of the transmission case to remove dirt and metal particles worn from moving parts. Then clean and check all parts for wear or damage.

As you inspect the transmission, look for:

● Gears with uneven wear, broken or chipped teeth, cracks or scoring.

● Badly worn or loose bearings and bearing rollers that are chipped or damaged.

● Excessive clearance between gears.

● Loose or damaged keys and keyways.

● Broken or distorted shifter detent springs and scored or flattened detent balls.

● Damaged or plugged transmission oil lines or passages in shafts.

● Badly worn splines on drive shafts and in sliding gear transmissions.

● Excessively worn teeth in shifter collars and hubs in collar shift transmissions.

Fig. 41 — Removing Differential Drive Shaft From Transmission Case

When reassembling the transmission, coat parts with clean new transmission oil of the type recommended by the manufacturer. Use of different lubricants could contaminate the new oil and reduce transmission life.

Follow the reassembly procedures outlined in the technical service manual. They can save time and help avoid mistakes which could lead to another transmission failure.

TROUBLESHOOTING BELT-TYPE TRANSMISSIONS

Problems in belt-drive transmissions are often similar to those found in regular belt drives, plus some additional troubles. If a belt-type transmission fails to function properly, check the points noted below and see "Troubleshooting Belt Drives" in Chapter 2. Refer to the technical service manual for repair or adjustment procedures and specifications.

Trouble	Possible Cause
Excessive belt wear	1. Weak belt tightener spring
	2. Worn, bent, or nicked sheaves
	3. Dirt in sheave grooves
	4. Belt contacting other machine parts
Belt slips under load	1. Control linkage out of adjustment
	2. Belt(s) badly worn or stretched
	3. Oil or grease on belt
	4. Overloading machine
	5. Using wrong belt
Difficult to change speeds	1. Linkage out of adjustment
	2. Variator sheave sticking or damaged
	3. Dirt in variator sheave
	4. Linkage needs lubrication

TROUBLESHOOTING GEAR-TYPE TRANSMISSIONS

The type of problem encountered in a transmission often tells what is causing the trouble. As shown below, there are several possible causes for each trouble, and each should be checked because more than one thing could be wrong at the same time.

Trouble	Possible Cause
Transmission noisy in neutral	1. Transmission not aligned with engine
	2. Bearings dry, badly worn, or broken
	3. Transmission oil level low
	4. Gears worn, scuffed, or broken
	5. Countershaft sprung or badly worn
	6. Excessive endplay of countershaft
Transmission noisy while in gear	1. Same causes noted in "Transmission noisy in neutral"
	2. Main shaft rear bearing worn or broken
	3. Engine vibration dampener defective
	4. Speedometer drive gears worn
	5. Clutch friction disk defective
	6. Gears loose on main shaft

Transmission hard to shift	1. Engine clutch not releasing
	2. Sliding gear tight on shaft splines
	3. Shift linkage out of adjustment
	4. Main shaft splines distorted
	5. Synchronizer damaged
	6. Sliding gear teeth damaged
Gears clash when shifting	1. Clutch not releasing
	2. Synchronizer defective
	3. Gears sticking on main shaft
Transmission sticks in gear	1. Clutch not releasing
	2. Detent balls stuck
	3. Shift linkage out of adjustment or not lubricated
	4. Sliding gears tight on shaft splines
Transmission slips out of gear	1. Shift linkage out of adjustment
	2. Gear loose on shaft
	3. Gear teeth worn
	4. Excessive endplay in gears
	5. Lack of spring tension on shift lever detent
	6. Badly worn transmission bearings
Transmission leaks oil	1. Oil level too high
	2. Gaskets damaged or missing
	3. Oil seals damaged or improperly installed
	4. Oil throw rings damaged, improperly installed, or missing
	5. Drain plug loose
	6. Transmission case bolts loose, missing, or threads stripped
	7. Transmission case cracked
	8. Lubricant foaming excessively

CHAPTER 5 REVIEW

1. Name two types of transmissions which use V-belts.

2. (Choose one) In a variable speed drive with input sheave of constant size, output speed (increases, decreases) as the belt pulls deeper into the variator sheave.

3. (Choose one) A friction disk drive can be used to control output (speed, direction, both).

4. Name four kinds of gear-type transmissions.

5. (Choose one) At high engine rpm, the drive belt runs in the (smallest, largest) diameter of the drive clutch.

6. Name two basic types of sliding gear transmissions.

7. Name a major advantage of constantly-meshing gears in a collar shift transmission.

8. (True or False) Synchronizer shaft transmissions do not allow for shift on-the-go.

9. How are speeds changed in a gear-type transmission?

CHAPTER 6

HYDROSTATIC TRANSMISSIONS

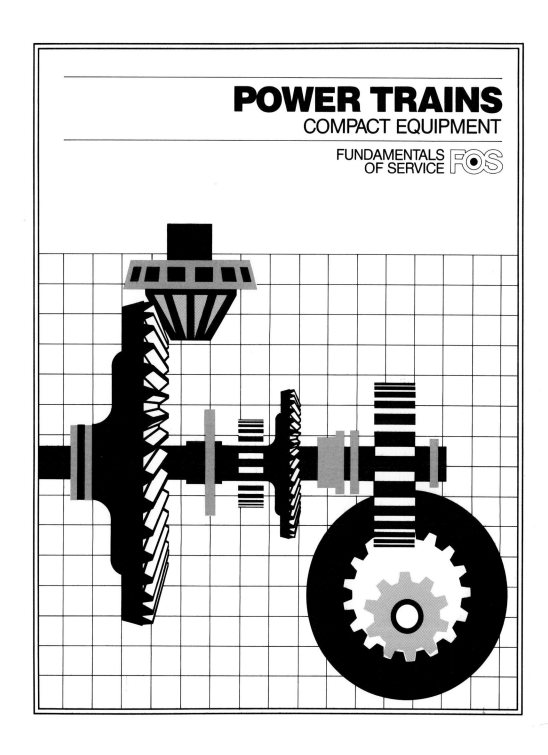

POWER TRAINS
COMPACT EQUIPMENT

FUNDAMENTALS
OF SERVICE FOS

SKILLS AND KNOWLEDGE

This chapter contains basic information that will help you gain the necessary subject knowledge required of a service technician. With application of this knowledge and hands-on practice, you should learn the following:

- **Theory of operation of the hydrostatic transmission.**

- **Types of hydrostatic drives used.**

- **Application of various types of hydrostatic drives.**

- **How a hydrostatic transmission works in forward, neutral, and reverse.**

- **Operation of a ball-type hydrostatic transmission.**

- **Advantages of hydrostatic drives.**

- **General repair and maintenance.**

- **General disassembly procedures for hydrostatic drives.**

- **General troubleshooting procedures.**

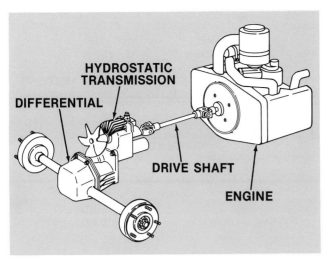

Fig. 1 — Hydrostatic Drive In A Complete Power Train

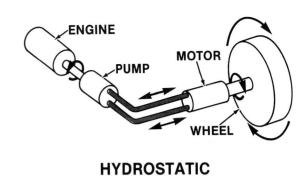

HYDROSTATIC

Fig. 2 — Hydrostatic Drives Use High Pressure And Low Velocity Fluid

INTRODUCTION

Hydrostatic transmissions (or drives) use fluids under pressure to transmit power from the engine to the drive wheels of the machine.

In a typical hydrostatic drive (Fig. 1), rotary power from the engine is turned into hydraulic power by a pump coupled directly to the engine crankshaft. A motor then converts hydraulic power back to mechanical rotary power for the drive wheels.

Because the flow of hydraulic power is easily started, stopped, and controlled, a hydrostatic drive can replace the regular clutch and transmission in the power train as shown in Fig. 1. Infinite speed and torque ranges from full forward to full reverse can be supplied by a hydrostatic drive. This permits precise control of machine travel speed without affecting the speed of auxiliary equipment such as mowers, sprayers, etc.

HYDROSTATIC TRANSMISSION OPERATION

Hydrostatic drives use **high pressure** hydraulic oil which is pumped at relatively **slow speeds** to transmit power from the engine to drive wheels. Do not confuse hydrostatic with hydrodynamic or torque converter drives which use **high speed** oil pumped at **low pressure.** (See Chapter 1 for for a description of hydrodynamic drives.)

In a hydrostatic drive (Fig. 2), energy is transferred by the fluid in a closed circuit between the pump and motor. The fluid, shown in red, moves through the system from pump to hydraulic motor. The rise in fluid pressure is produced by the motor's partial resistance to turning.

THEORY OF OPERATION

If two cylinders (Fig. 3), each containing a piston, are connected by a hydraulic line, the line and one cylinder are filled with oil, and a force (from the engine) is applied to the left piston, that left piston will force oil from the left cylinder, through the line, and into the right cylinder. The oil will not compress, so it forces the right piston out with a force equal to that applied to the left piston.

In a hydrostatic drive, several pistons work together in groups to transmit power. One group of pistons in the pump sends power to another group of pistons in the motor.

Each group of pistons is in a cylinder block (Fig. 4). The pump cylinder block revolves around the input (drive) shaft, and the motor cylinder block revolves around the output (driven) shaft. The pistons also move back and forth in cylinders in the blocks.

Fig. 3 — Fluids Transmit Force

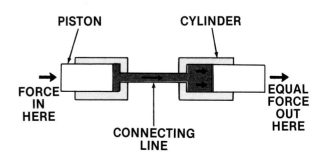

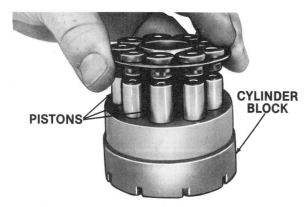

Fig. 4 — Hydrostatic Drive Cylinder Block And Pistons

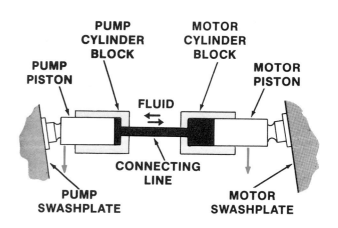

Fig. 5 — Two Connected Cylinders With Swashplates

As the cylinder block turns, the pistons, in both the pump and motor, ride against the surfaces of the swashplates, moving in and out, following the angle of the plate's surface. This provides a pumping action. The angle of the swashplates (Fig. 5) can be varied so the volume and pressure of oil pumped by the pistons can be changed or the direction of flow reversed.

A pump or motor with this kind of variable swashplate is called a variable-displacement unit. If the swashplate is fixed, it is called a fixed-displacement pump or motor. Fixed and variable-displacement pumps and motors can be used in any combination.

The motor swashplate in Fig. 6 is fixed (fixed-displacement), but the pump swashplate is movable (variable-displacement). As the pump cylinder block rotates, the pistons move around the sloping face of the swashplate, sliding in and out of their cylinder bores to pump oil out to the motor. The more the pump swashplate is tilted, the more oil it pumps (displacement) with each piston stroke and the faster the motor is driven.

Pressure results whenever fluid flow is resisted. The resistance may be caused by acceleration of the machine or a normal load.

As stated before, the motor swashplate (Fig. 6) is fixed so that the stroke of the motor pistons is always the same. Thus the speed of the motor rotation cannot be changed except as it is driven faster or slower by oil from the pump.

In this situation, a given volume of oil forced out of the pump causes the motor to turn at a given speed. More oil will increase motor speed; less oil will slow the motor.

The pump is driven on the machine's engine at the speed set by the operator. And it maintains a flow of high-pressure oil to the motor. Since the motor is linked to the machine drive wheels, oil flow to the motor determines machine travel speed.

Only three factors control operation of a hydrostatic drive:

● **Rate** of oil flow — gives the speed

● **Direction** of oil flow — gives the direction

● **Pressure** of oil — gives the power

These three factors can be varied precisely to give the speed and torque needed.

The heart of a hydrostatic drive is the pump-motor team, (Fig. 7) although there is also a reservoir to store oil, a filter to remove contaminants, and a cooler to remove excess heat from the oil.

Basically however, the pump and motor are joined in a closed loop; the return line from the motor is joined directly to the pump intake, rather than to the reservoir. The charge pump simply maintains an adequate supply of oil from the reservoir for the main pump.

TYPES OF HYDROSTATIC DRIVES

Displacement is the quantity of fluid which a pump can move or a motor can use during each revolution. It is directly related to the power output of the drive. Power is a combination of torque and speed.

As you have already learned, pumps and motors can have either fixed or variable displacement.

There are four possible pump-motor combinations:

● *Fixed-displacement pump driving a fixed-displacement motor*

● *Variable-displacement pump driving a fixed-displacement motor*

● *Fixed-displacement pump driving a variable-displacement motor*

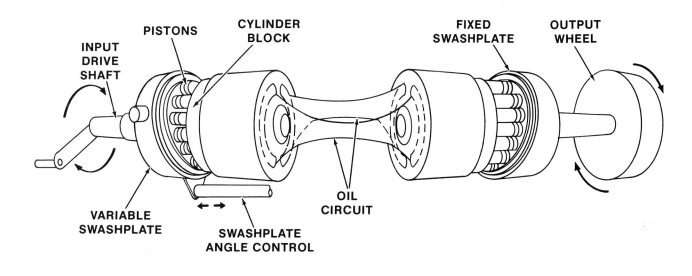

PUMP
(VARIABLE DISPLACEMENT)

MOTOR
(FIXED DISPLACEMENT)

Fig. 6 — Variable-Displacement Pump Driving Fixed-Displacement Motor

● *Variable-displacement pump driving a variable-displacement motor*

Just remember that for any of these combinations, the output power must equal the input power minus any negligible losses from friction, heat, minor leakage within the system, etc. Fig. 8 shows each combination.

The No. 1 circuit resembles a gear drive. Power is transmitted without altering the speed or power between the engine and the load. A constant input speed and torque gives a constant output power. If either speed or torque

is increased and the other held constant, power output increases.

The variable pump flow in No. 2 circuit results in variable speed output, but torque output is constant for any given pressure. This circuit gives variable speed and constant torque.

Output speed in No. 3 circuit is varied by changing the motor displacement. But, with constant power input, if the motor displacement decreases, output speed increases and output torque drops.

Fig. 7 — Complete "Closed Hydraulic Loop" System For A Hydrostatic Drive

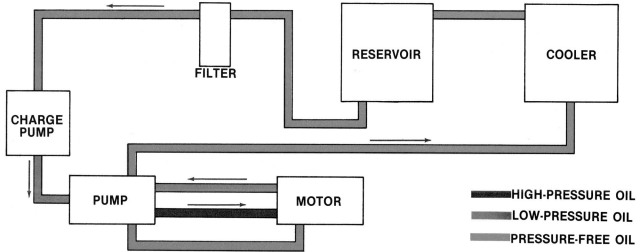

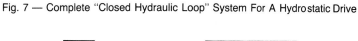

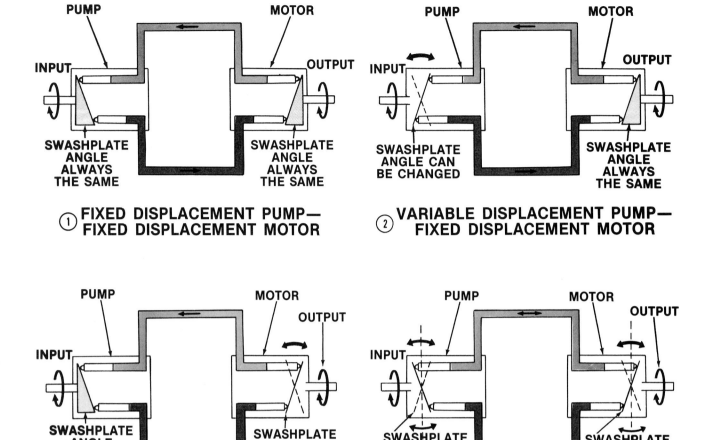

Fig. 8 — Pump-Motor Combinations For Hydrostatic Drives

No. 4 circuit is the most flexible and expensive of all the circuits and the most difficult to control. It is capable of operating like any of the other three circuits.

REVERSING IN VARIABLE PUMP-MOTOR CIRCUITS

Output shaft rotation can be reversed in variable circuits by shifting either the pump or the motor swashplates over center (Fig. 9).

● In neutral, the swashplate is vertical and no oil is pumped.

● In forward, the swashplate is tilted and oil is pumped as shown at the top.

● In reverse, the swashplate is tilted **the opposite way** and oil is pumped the opposite direction.

APPLICATION OF HYDROSTATIC DRIVES

Due to their versatility and operating convenience, hydrostatic drives are being used on more and more equipment. They are also found in several different pump-motor configurations. And different ways are used to transfer power from the hydrostatic motor to the machine drive wheel. Common possibilities are:

● *One pump and one motor.*

● *One pump and multiple motors.*

● *One pump and one motor for each drive unit.*

● *Hydrostatic drive coupled with mechanical transmission.*

● *Hydrostatic drive coupled to differential and final drive.*

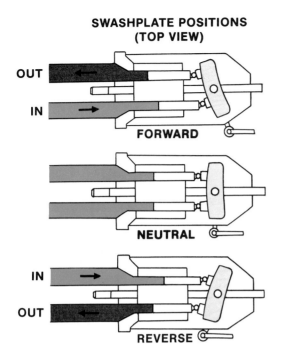

SWASHPLATE POSITIONS
(TOP VIEW)

OUT

IN

FORWARD

NEUTRAL

IN

OUT

REVERSE

Fig. 9 — Reversing Rotation With Variable Displacement Pumps Or Motors

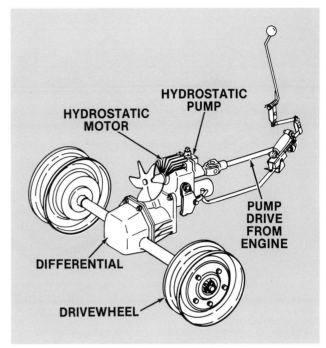

Fig. 10 — Hydrostatic Drive With One Pump And One Motor

ONE PUMP AND ONE MOTOR

In the most common application, a single pump drives one motor (Fig. 10) which is connected to the drive wheels through a conventional differential and final drive.

In a one pump-one motor drive, the two components may be assembled into a single unit (Fig. 10) with oil passages machined into the coupling between pump and motor. This provides a very compact system for smaller machines such as lawn and garden tractors.

In other machines, the pump and motor are mounted separately and usually connected by hydraulic hoses (Fig. 11). (Hoses can absorb shock loadings better than rigid metal tubing if the drive is suddenly shifted between extreme high and low speeds). This arrangement also permits designers to make better use of space, or put the motor where it would be impractical to run a mechanical drive train.

ONE PUMP AND MULTIPLE MOTORS

A separate hydrostatic drive motor is installed directly in the hub of each drive wheel on some machines (Fig. 12). In such designs, the hydrostatic drive and hydraulic lines completely replace the mechanical power train — clutch, transmission, differential, and final drive. Planetary gears in each wheel hub may be used to reduce motor speed to the correct level for machine operation.

The flow control valve automatically provides differential action to the wheel motors for easier turning.

MULTIPLE PUMPS AND MOTORS

Controlling power individually to each side of a machine also permits the drive system to steer the wheels. In the all-terrain mower (Fig. 13), a tandem pump provides separate streams of oil to drive wheels on the right and on the left. A hydrostatic motor is mounted in the hub of each drive wheel.

Fig. 11 — Hydrostatic Drive With One Pump And Remote Motor

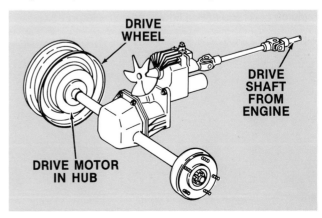

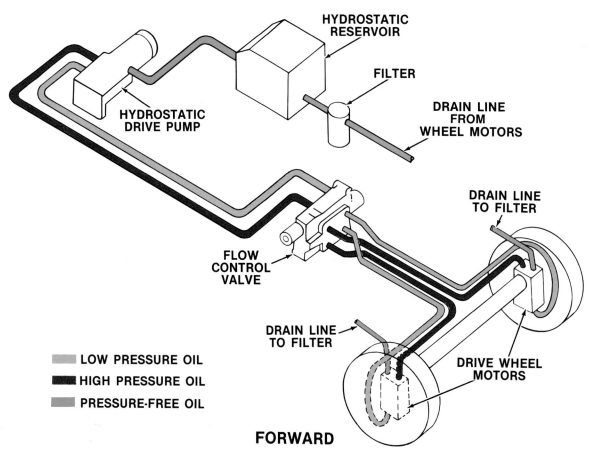

FORWARD

Fig. 12 — Hydrostatic Drive With One Pump And Multiple Motors

Manipulating steering control levers controls oil flow to the wheel motors. Directing equal amounts of oil to each side causes the machine to move straight ahead or reverse. However, directing more oil to motors on one side than to the other causes the machine to turn away from the faster side. Reversing the wheels on one side and powering the other wheels forward permits the machine to pivot in its own tracks.

Similar maneuverability can be obtained with one or two drive wheels (and motors) on each side of the machine and either one pump for each side or separate controls to divide and regulate oil flow from a single pump.

HYDROSTATIC DRIVE COUPLED WITH MECHANICAL TRANSMISSION

To provide the proper combination of speed and power for particular jobs, some machines combine a hydrostatic drive with a mechanical transmission (Fig. 14). In this case, the output shaft of the hydrostatic motor is connected directly to a two-speed, sliding-gear transmission that is built into the differential housing. This permits changing speeds to suit working conditions.

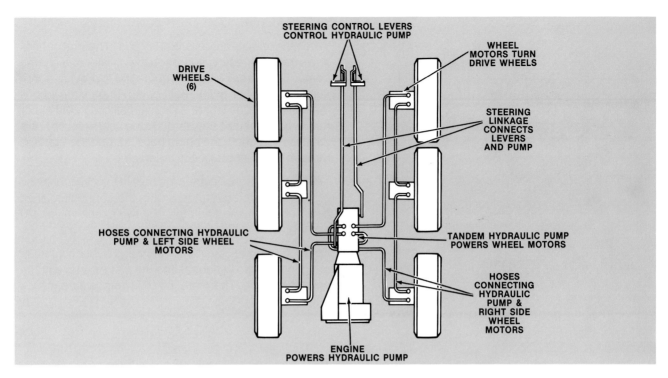

Fig. 13 — Hydrostatic Drive With Tandem Pump And Multiple Motors

OPERATION OF HYDROSTATIC DRIVES

To see how a typical hydrostatic drive works, we will look at the combination found in most hydrostatic drives, an axial piston pump (Fig. 15) and motor (Fig. 16).

The pump has a variable displacement while motor displacement is fixed. This combination is also shown in Fig. 6.

An explanation of how the complete system operates in neutral, forward, and reverse follows.

Fig. 14 — Hydrostatic Drive Coupled With Mechanical Transmission

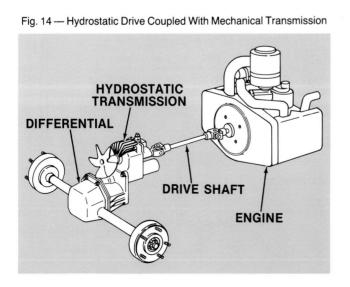

Fig. 15 — Axial Piston Variable-Displacement Pump

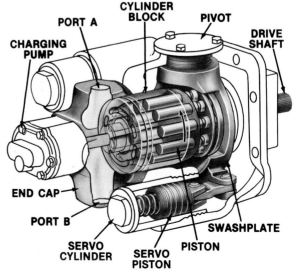

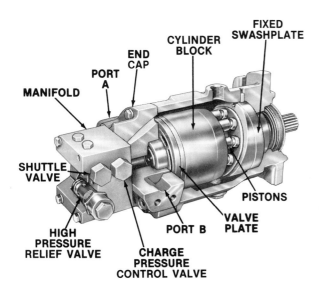

Fig. 16 — Axial Piston Fixed-Displacement Motor

NEUTRAL (FIG. 17)

When the speed range control lever is in neutral, free oil flows from the reservoir through the oil filter to a charge pump. This pump (which is a gear-type to provide a constant supply of oil to the main pump) pumps the oil into the center section where it circulates through the motor housing. Oil then passes through the center section, circulates throughout the pump housing, and then returns through the oil cooler to the reservoir.

Trapped oil (shown in green) is held in the cylinder block of the pump, in the motor, and in the connecting lines between the pump and motor by two check valves (A) and (B) in the center section.

When the control lever is in neutral, the swashplate in the pump is also in neutral and the pistons in the pump are not stroking. Therefore, no oil is being pumped.

Fig. 17 — Hydrostatic Drive in Neutral

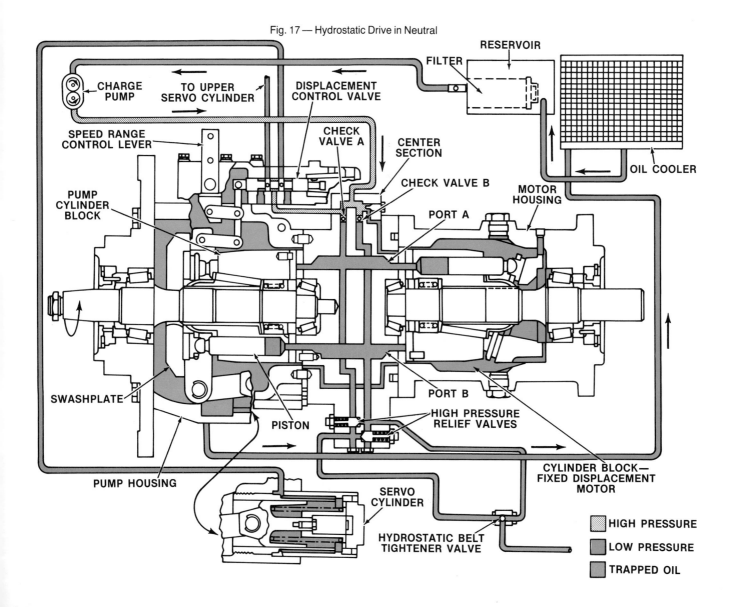

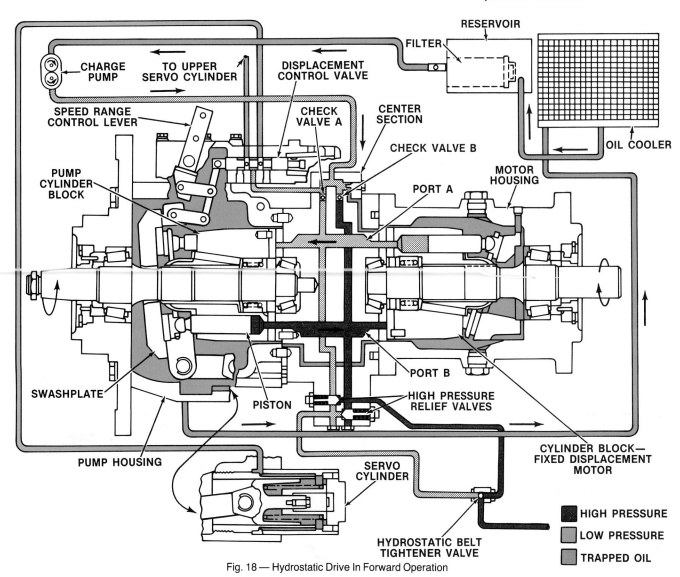

Fig. 18 — Hydrostatic Drive In Forward Operation

The cylinder block in the pump, driven by the machine engines, rotates in a counterclockwise direction as viewed from the input end of the pump.

Because oil is not being pumped to the motor, the cylinder block in the motor is stationary and the output shaft does not move.

The hydrostatic belt tightener valve provides for pressure oil to assist a spring to keep proper tension on the belt.

FORWARD (FIG. 18)

When the speed range control lever is moved forward, the spool in the displacement control valve moves out of neutral. This allows oil to flow into the lower servo cylinder forcing the swashplate (which is connected to the servo cylinder) to tilt. Oil expelled by the opposing servo

cylinder (not shown) returns through the displacement control valve to the pump case.

When the swashplate reaches the tilt set by the control lever, the displacement control spool returns to neutral (as shown in Fig. 18). This traps the oil to both servo cylinders and holds the swashplate tilted as shown. The swashplate will remain in this position until the speed control lever is moved by the operator.

With the pump drive shaft and cylinder block rotating counterclockwise and the top of the swashplate tilted to the left, low pressure oil enters port A through check valve A. This forces the pistons in the pump, that align with port A, to go out against the swashplate, filling the pump cylinder block. The further the swashplate is tilted, the more oil the pump receives. As the driveshaft continues to turn the cylinder block, the pistons are forced back into the block because of the tilted position of the swashplate. As the pistons move back into the block, oil in the cylinder becomes pressurized. Then, when the cylinder is aligned with port B, pressure oil moves through port B into the cylinder in the motor that is aligned with port B.

The distance the pistons in the pump reciprocate in and out of the cylinder block depends on the angle the pump swashplate is tilted. This determines the volume of oil displaced per revolution of the pump. The greater the angle, the greater the volume and the more oil flows from the pump. Thus, as the swashplate angle is changed, the volume of oil displaced from the pump will also change. The greater the angle, the greater the volume, hence a greater flow of oil from the pump.

As pressure oil moves from pump to motor through port B, pistons in cylinders in the motor that align with port B are pushed against the fixed swashplate. The pistons slide down the inclined surface and resulting forces cause the cylinder block to rotate which in turn rotates the output shaft. This drives the differential and transmission causing the machine to move forward.

The cylinder block in the motor rotates in a clockwise direction (viewed from the output shaft end of the motor) when traveling forward. As the cylinder block continues to rotate, oil is forced out of port A, through the center section, to the pump.

The oil returned by the motor to the pump through the center section is recirculated through the pump cylinder block assembly and back to the motor. This is called a "closed circuit" system in that the oil continually circulates between the pump and the motor. The only oil introduced into the closed circuit from the reservoir comes from the charge pump.

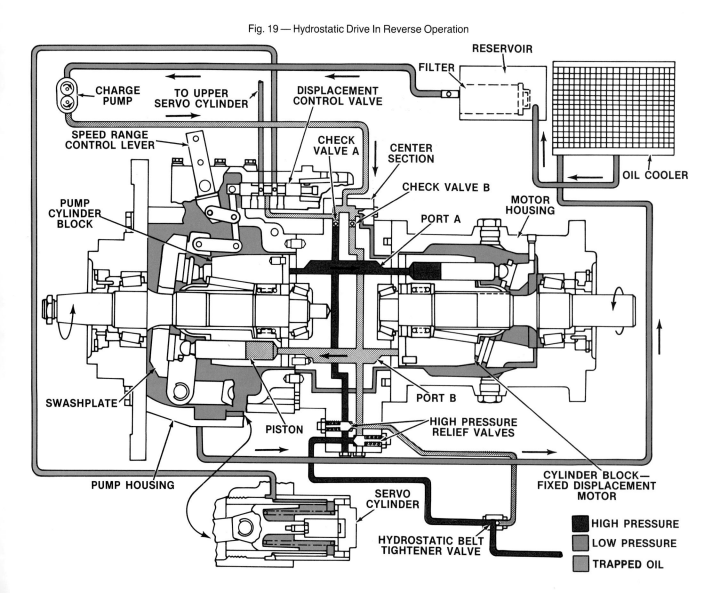

Fig. 19 — Hydrostatic Drive In Reverse Operation

If pressure exceeds the rated pressure of the high pressure relief valves, the relief valve opens and oil by passes the cylinder block assembly in the motor, slowing or stopping the machine. The bypassing oil returns to the pump.

REVERSE (FIG. 19)

When the speed range control lever is moved to reverse, the spool in the displacement control valve moves out of neutral. This allows pressure oil to flow in the upper servo cylinder, tilting the swashplate as shown.

When the swashplate reaches its desired tilt, as set by the control lever, the displacement control spool returns to neutral (as shown in Fig. 19). This traps the oil to both servo cylinders and holds the swashplate tilted as shown. The swashplate remains in this position until the speed control lever is moved again by the operator.

With the pump drive shaft and cylinder block rotating counterclockwise and the top of the swashplate tilted to the right, port B becomes the *inlet* and port A the *outlet*

(high pressure port). As the cylinder rotates past port B, check valve B opens and oil is forced by the charge pump into the piston bores, which align with port B. Then as rotation continues, the oil is pressurized and forced out of port A by each of the pistons as they align with port A. This oil flows to the motor.

As high pressure oil from the pump enters port A of the motor, it pushes the motor pistons against the swashplate. The pistons slide down the inclined surface of the swashplate, rotating the cylinder block which in turn rotates the drive shaft, driving the machine — but in reverse.

As the motor cylinder block continues to rotate, oil is forced out port B at low pressure and returns to the pump.

*NOTE: In the system shown here, the **pump** drive shaft and cylinder block always rotate counterclockwise, but the **motor** drive shaft and cylinder block can rotate in either direction, depending on the direction of oil entering from the pump.*

Fig. 20 — Integral Pump-Motor For Hydrostatic Drive

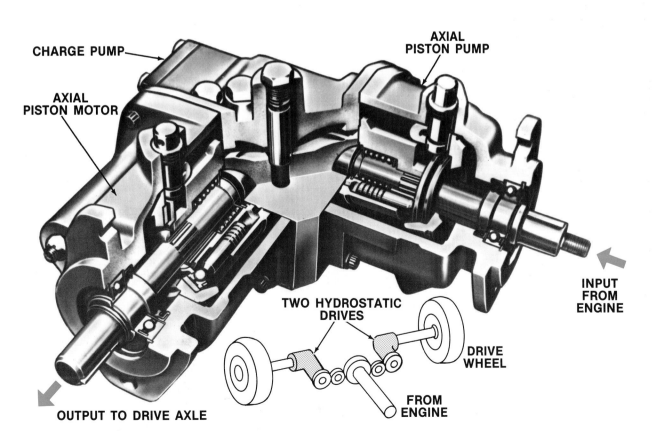

A TYPICAL SYSTEM

In the system shown (Fig. 20), an integral pump-motor unit is used on each drive axle. The pump and motor are connected at right angles and closed circuit oil paths between the pump and motor are built into the housing.

Both pump and motor have variable displacement, providing variable torque and speed.

A charge pump as well as pressure control and relief valves are built into the system. The charge pump is a vane type with fixed displacement.

The swashplates are controlled through linkage fastened to yokes above the pump and motor. Fig. 21 illustrates how it works.

Oil flows from the reservoir (Fig. 21) to the inlet port of the charge pump. Oil then passes from the charge pump to the control valve. If the valve is not actuated (open for passage of oil), oil passes through the control valve into a common line with oil from the other control valve, back to the filter. After oil is filtered and cooled, it enters the valve block inlet port of both drives.

The valve block has common oil passages to permit oil to flow from the piston pump to the piston motor and return to the piston pump. This is a closed circuit. When any leakage occurs, the circuit is replenished by cooled oil entering the valve block.

The replenishing oil overcomes the 10 psi (69 kPa) spring setting and discharges into the inlet (low pressure) side of the piston pump at a pressure greater than

10 psi (69 kPa). Any excess pressure in the common passage empties over the 25 psi (172 kPa) relief valve into the motor housing and back to the reservoir. The 25 psi (172 kPa) relief valve consists of a sleeve poppet and spring. Pressure acts directly on the poppet against the spring.

Relief valve action is provided to accommodate pressure surges in the output passages. The replenishing and relief valve cartridges are pre-set.

The relief valve operates when the outlet pressure on the end of the inner poppet creates a force greater than the pilot spring force of 4,500 psi (31,025 kPa) and unseats the inner poppet. This reduces pressure in the spring chamber until the outer poppet moves, discharging excess pressure to the relief cavity. The inner poppet is held open by its pin until the system pressure drops below 4,500 psi (31,025 kPa).

Since either port of the piston pump can be inlet or outlet, two replenishing and two relief valves are required to provide proper relief protection.

The pump for each side is driven by the engine, and each motor drives the wheel on that side. The machine is steered by actuating the pump swashplate on each axle. For example, a steering wheel can be linked indirectly to the swashplates for gradual turns. But, for sharp turns or reversing an axle, steering levers — one for each side — can be used.

Fig. 21 — Oil Flow In Hydrostatic Drive Axle

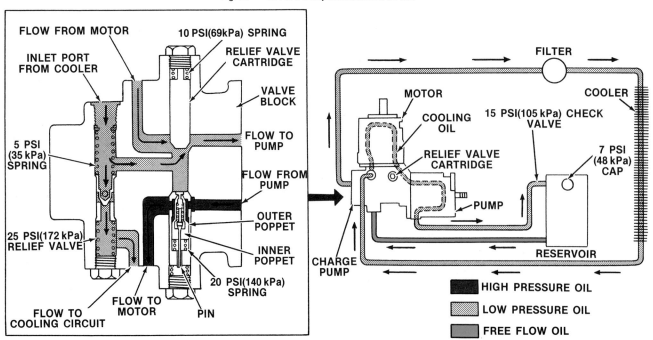

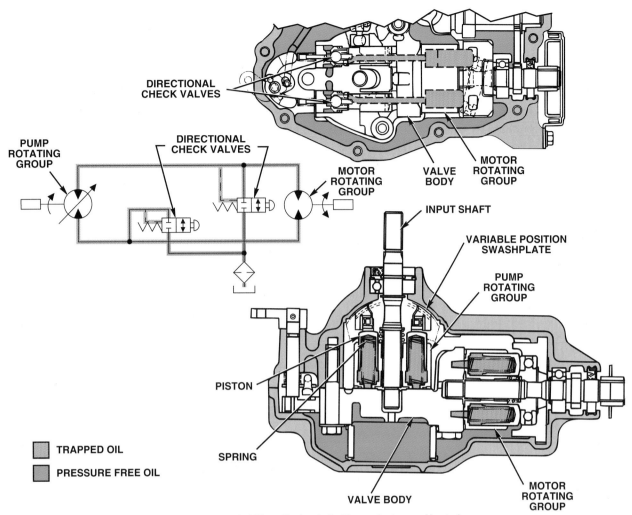

DIRECTIONAL
CHECK VALVES

PUMP
ROTATING
GROUP

DIRECTIONAL
CHECK VALVES

MOTOR
ROTATING
GROUP

VALVE
BODY

MOTOR
ROTATING
GROUP

INPUT SHAFT

VARIABLE POSITION
SWASHPLATE

PUMP
ROTATING
GROUP

PISTON

SPRING

TRAPPED OIL

PRESSURE FREE OIL

VALVE BODY

MOTOR
ROTATING
GROUP

Fig. 22 — Ball-Type Hydrostatic Transmission — Neutral

To reverse either side, the pump swashplate is moved overcenter, reversing the flow circuit which drives that motor.

The ball-type transmission is a type of hydrostatic transmission. Let's look at its operation.

BALL-TYPE HYDROSTATIC TRANSMISSION

Neutral (Fig. 22)

When the transmission is in neutral with the engine running, the input shaft and the pump rotating group are turning. Springs force the pump pistons against the variable position swashplate which is parallel to the pump body.

With the swashplate parallel to the pump body, there is no reciprocating motion of the pistons; therefore, no oil is pumped to the motor rotating group. Both directional check valves are closed because there is no demand for oil in the pump.

Oil in the valve body and the motor rotating group is trapped, because oil cannot be displaced to the pump. In order to receive oil, the pistons in the pump would have to reciprocate. Tractor position remains stationary due to dynamic braking action within the closed loop.

Forward (Fig. 23)

With the engine running, the input shaft and pump rotating group are turning. When the directional control lever is moved to the forward position this angles the variable position swashplate within the transaxle.

Springs inside the piston bores force them against the swashplate. As the pistons follow the contour

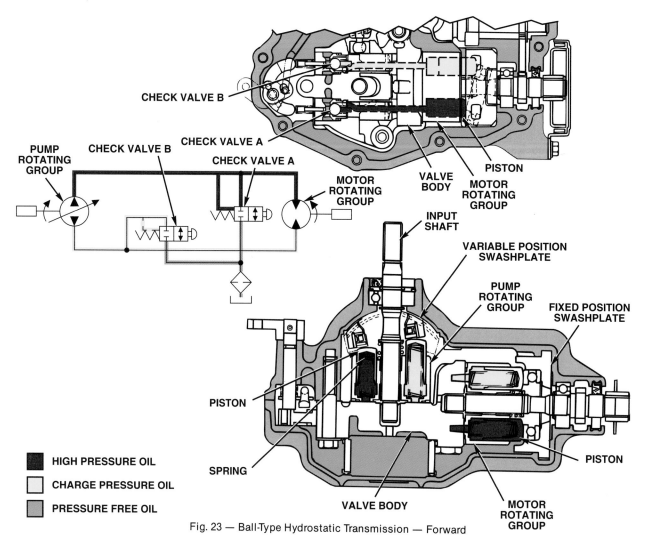

CHECK VALVE B

CHECK VALVE A

PUMP
ROTATING
GROUP

CHECK VALVE B

CHECK VALVE A

CHECK VALVE A

MOTOR
ROTATING
GROUP

VALVE
BODY

PISTON

MOTOR
ROTATING
GROUP

INPUT
SHAFT

VARIABLE POSITION
SWASHPLATE

PUMP
ROTATING
GROUP

FIXED POSITION
SWASHPLATE

PISTON

PISTON

■ HIGH PRESSURE OIL

□ CHARGE PRESSURE OIL

▨ PRESSURE FREE OIL

SPRING

VALVE BODY

MOTOR
ROTATING
GROUP

Fig. 23 — Ball-Type Hydrostatic Transmission — Forward

of the swashplate they move outward drawing oil in their bores. As the pistons continue to rotate, swashplate angle forces them back in their bores and displaces the oil through porting in the valve body.

The oil displaced by the pump, travels through one side of the valve body to the motor rotating group. The motor rotating group works in conjunction with a fixed position swashplate. The oil enters the piston bore through a port in the valve body at a point where the piston is compressed in its bore.

As the oil fills the piston bore, the piston is forced out and follows the contour of the swashplate. This creates the rotary motion in the motor group for forward drive. Oil pressure within these components is directly proportional to the load encountered. This is known as the high pressure side.

As the motor continues to rotate, the piston is now

compressed by the angle of the swashplate and oil is forced from the piston bore into the other port within the valve body. This oil is returned to the motor to be drawn into a piston bore of the pump rotating group. There is minimal oil pressure from the pump back to the motor and this is referred to as the low pressure side of the system.

This cycle where oil is moved from the pump to the motor and back to the pump again is referred to as the closed loop.

A certain amount of oil is lost in the closed loop due to leakage, so there needs to be a means of replacing the oil. This is accomplished by the directional check valves.

When in forward direction, check valve (A) is forced closed by the pressure on this side of the closed loop. A vacuum created by the pump on the low pressure side of the closed loop will open check

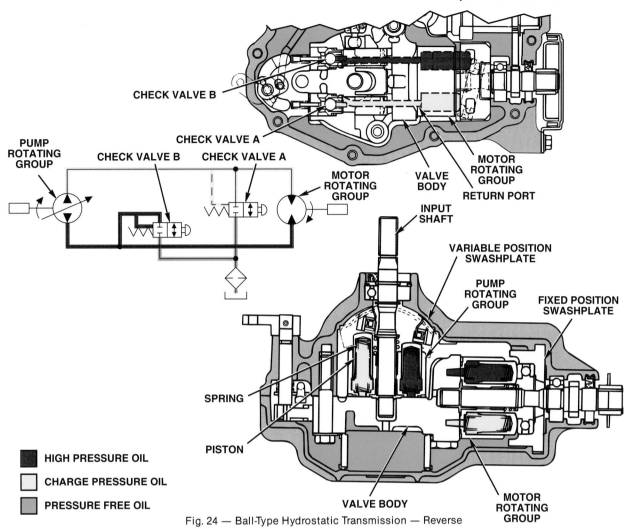

CHECK VALVE B

PUMP ROTATING GROUP

CHECK VALVE A

CHECK VALVE B   CHECK VALVE A

MOTOR ROTATING GROUP

VALVE BODY

MOTOR ROTATING GROUP

RETURN PORT

INPUT SHAFT

VARIABLE POSITION SWASHPLATE

PUMP ROTATING GROUP

FIXED POSITION SWASHPLATE

SPRING

PISTON

HIGH PRESSURE OIL

CHARGE PRESSURE OIL

PRESSURE FREE OIL

VALVE BODY

MOTOR ROTATING GROUP

Fig. 24 — Ball-Type Hydrostatic Transmission — Reverse

valve (B) and allow oil into the suction side of the pump to make up oil lost to leakage.

Reverse (Fig. 24)

With the engine rotating, the input shaft and the pump rotating group are turning. When the directional control lever is moved to the reverse position, this angles the variable position swashplate within the transaxle. You will note that this angle is opposite that of forward direction which changes the position where the pistons draw in oil.

Springs inside the piston bores force them against the swashplate. The angle of the swashplate causes the pistons on one side of the rotating group to be compressed within their bores and pistons on the other side to move outward in their bores.

As the rotating group turns, the pistons follow the contour of the swashplate moving in and out. This action draws oil into the piston bore as it moves out and forces the oil out as the piston moves back in.

The oil being displaced from the pump travels through one side of the valve body to a port that lines up with the piston bores of the motor rotating group. The oil enters the piston bore at a point where the piston is compressed. This causes the piston to expand in the bore. In order to move out in its bore the piston has to follow the contour of the fixed position swashplate. This causes the motor rotating group to turn.

The piston continues to expand in its bore and follow swashplate angle until it no longer is lined up with the delivery port of the valve body. Since oil is no longer being displaced to that piston, it no longer adds to the rotory motion of the motor. But, pistons behind it are now filling with oil, so the motion continues.

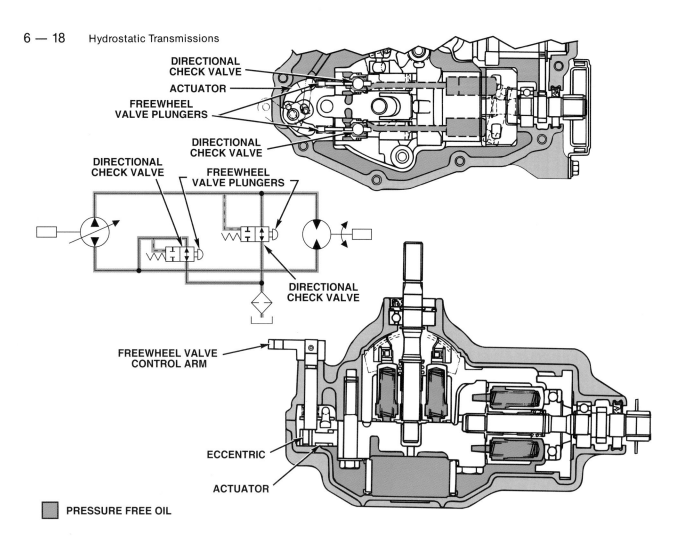

DIRECTIONAL CHECK VALVE

ACTUATOR

FREEWHEEL VALVE PLUNGERS

DIRECTIONAL CHECK VALVE

DIRECTIONAL CHECK VALVE

FREEWHEEL VALVE PLUNGERS

DIRECTIONAL CHECK VALVE

FREEWHEEL VALVE CONTROL ARM

ECCENTRIC

ACTUATOR

PRESSURE FREE OIL

Fig. 25 — Ball-Type Hydrostatic Transmission — Freewheel

As the motor continues to revolve, the piston is forced back in its bore by the swashplate. As this is happening, the piston is in alignment with the return port in the valve body and oil is directed back to the suction side of the pump.

The oil path from the pump to the motor is known as the high pressure side because oil pressure builds depending on system load. The oil path from the motor back to the pump is known as the low pressure side. From forward drive to reverse these two paths switch which gives opposite rotation of the motor.

A certain amount of oil is lost from the pump and motor rotating groups. This is due to normal leakage or may be caused by component wear. Since this oil is recirculated, there needs to be a means of

replacing it. This is accomplished by the directional check valves.

When in reverse, check valve (B) is closed due to the high pressure forcing it against its seat. Check valve (A) will open if the pump demands oil on the low pressure side of the closed loop.

Freewheel (Fig. 25)

When the freewheel valve control arm is engaged, this rotates the eccentric that moves the actuator. The actuator moves the freewheel valve plungers against the directional check valves. This forces the check valves off their seats and allows a path for oil to flow from both sides of the motor to the reservoir. Normally, the motor would have excessive resistance to movement due to the dynamic braking of the hydraulic closed loop.

ADVANTAGES OF HYDROSTATIC DRIVES

Hydrostatic drives usually cost more than mechanical transmissions, but they offer several advantages such as:

- *Variable speeds and torques.*

- *Easy one-lever control.*

- *Smooth shifting without "steps".*

- *Shifts "on the go".*

- *High torque available for starting up.*

- *Flexible location — no drive lines.*

- *Compact size.*

- *Elimination of clutches and large gear trains.*

- *Reduction of shock loads.*

- *Low maintenance and service.*

REPAIR AND MAINTENANCE OF HYDROSTATIC DRIVES

Hydrostatic drives are fairly easy to maintain. The fluid lubricates system parts. Filters remove impurities. The oil cooler helps prevent overheating. And relief valves protect the system from overloading.

But like any other mechanism, hydrostatic drives must be properly operated and maintained. A hydrostatic drive can be damaged by too much speed, heat, pressure, or contamination.

It is possible on some machines to disengage the hydrostatic pump from the engine to permit easier starting in cold weather. Also, and **this is very important,** to prevent hydrostatic system damage during cold weather, run the engine long enough to warm up the oil before moving hydrostatic control levers from the neutral position. This may take 10-15 minutes during cold weather. But, cold oil can cause charge pump cavitation which can damage the piston pump. Install an oil heater that plugs into a standard 110 volt electrical outlet if necessary to reduce warmup time.

Keep the hydrostatic system clean! Engineers and service technicians agree that dirt and water are the biggest enemies of any hydrostatic drive or hydraulic system. Dirt causes scratching and rapid wear between close fitting parts. And water causes rusting or combines with contaminants in the oil to form corrosive acids which damage parts.

Oil and filters must be changed at the recommended intervals — or more frequently when operating under adverse conditions. Be certain to use the oil and filters recommended by the machine manufacturer. Other oils and filters may be incompatible with system parts, seals or operating requirements.

DO IT RIGHT

Before removing any part of the hydrostatic system to check or add oil, drain oil, replace a filter, or make repairs, clean that entire area of the machine so dirt can't accidentally enter.

If the system must be repaired, use a steam cleaner or high pressure washer to clean the pump and motor before removing any parts. But, don't get water into the system. Make sure all hoses and connections are tight before spraying the units.

If steam cleaning or pressure washing is not possible, use fuel oil or a suitable solvent to remove dirt and grease. Do not use paint thinner, acetone, or gasoline.

Do not attempt to disassemble or repair a hydrostatic pump, motor, or control system unless you have had adequate training, have the proper tools available, and have a clean area in which to work.

Always use a clean work bench or table when disassembling the pump or motor for servicing. And never perform internal service work on the shop floor or on the ground, or where there is danger of dust or dirt being blown into parts. Also be sure all tools are clean and free of grease or dirt.

CHANGING HYDROSTATIC TRANSMISSION OIL

Refer to the applicable operator's manual for recommended oil change intervals and specific procedures for your particular tractor.

When changing oil, elevate the front of the tractor slightly and remove the drain plug (Fig. 26). Remove filter and clean the filter sealing area thoroughly. After all oil has drained, replace drain plug. Fill the new filter with oil, install, and hand-tighten only.

On some compact tractors the engine must be running at idle speed while filling with fresh oil. Again, check your specific operator or service manual. To fill, remove fill plug (Fig. 27) and add the proper amount of recommended oil. If your tractor is equipped with a sight tube (Fig. 27), add oil to the level specified in the operator's manual.

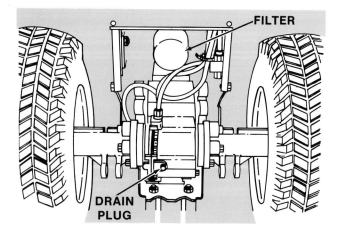

Fig. 26 — Changing Hydrostatic Drive Oil

DISASSEMBLY

Have these items available before disassembling the pump or motor:

• Several clean plastic plugs to seal openings when removing hydraulic lines or hoses.

• Several clean plastic bags to cover ends of lines, hoses, or openings. Secure bags with rubber bands or tape.

• A clean container of solvent to wash internal parts. Use clean compressed air to dry parts after washing. Use only clean lint-free cloths when wiping or drying parts.

• A clean container of transmission fluid to lubricate internal parts as they are assembled. (Do not use other types of oil as they may be incompatible with the system or recommended transmission oil.)

• A clean container of petroleum jelly (petrolatum) to lubricate surfaces where noted during reassembly.

When servicing the unit, take the following precautions:

• Seal any openings when doing service work to prevent dirt from entering the system. Be careful that all parts are clean before replacing them.

• Whenever units are serviced, always install new O-rings, seals, and gaskets during reassembly. This will provide tight seals for mating parts and help prevent leakage.

• Never service any part of the hydrostatic drive system with the machine engine running unless specifically instructed to do so in the operator's manual or technical service manual.

• Never operate the hydrostatic drive without oil, even for a very short time. Always check the oil supply before servicing the system. Low oil level could indicate specific types of problems within the system.

• Before storing the system, be sure the hydrostatic system is filled and the reservoir cap is tight to keep out moisture. The oil will prevent rust and corrosion.

• Always refer to the technical service manual for specific service and repair procedures.

Remove hydrostatic drive from vehicle as explained in the technical service manual. To remove the motor housing and pinion gear, remove the retaining ring (Fig. 28) and slide the pinion gear from the motor shaft. Remove four capscrews and lift motor housing and gasket from the transmission.

Fig. 27 — Filling Hydrostatic Transmission

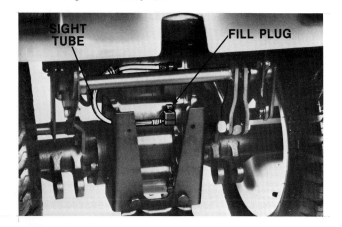

Fig. 28 — Removing Motor Housing And Pinion Gear

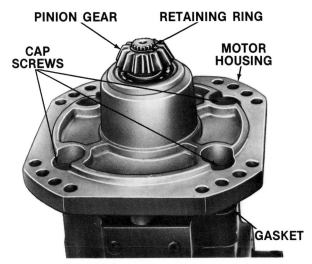

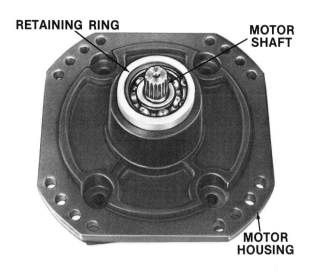

Fig. 29 — Removing Motor Shaft And Motor Housing From Hydrostatic Motor

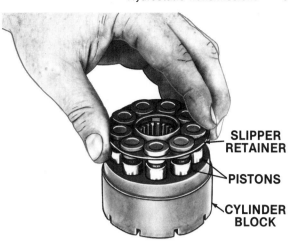

Fig. 31 — Inspect Pistons For Wear Or Damage

Remove the retaining ring (Fig. 29) to permit removal of the motor shaft and motor housing. Then remove the thrust plate, cylinder block, and motor valve plate (Fig. 30). Follow similar steps to remove the pump from the other end of the transmission.

Gently lift all pistons from the cylinder block (Fig. 31) and check for free movement of pistons in cylinder block bores. A scored or damaged piston will require replacement of the complete assembly as they are sold only in matched piston-cylinder block sets because of the close tolerances involved.

Carefully inspect all other parts of the pump and motor for wear, damage, or scratches which could cause oil leakage or damage to other parts.

NOTE: Most late model hydrostatic transmissions are not equipped with slippers on the pistons.

If the lubricant hole in the slipper (Fig. 32) is plugged, blow it clean with compressed air. But, if edges of slippers are too badly worn (rounded more than 1/32 inch for the unit shown), replace the entire cylinder block assembly. Avoid interchanging pistons between the pump and motor because each set is matched for correct fit in each unit. The differences cannot be detected by visual inspection, so keep parts separated while repairs are being made.

After all parts are thoroughly cleaned and inspected, replace any items which are worn or damaged and reassemble the transmission. Coat all parts with clean, new hydrostatic fluid for easier assembly and to provide some initial lubrication when the unit is restarted. Be

Fig. 30 — Removing Cylinder Block From Hydrostatic Motor

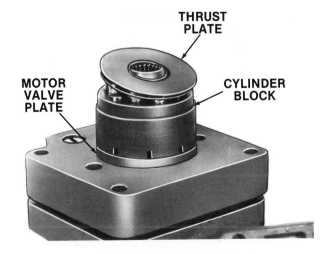

Fig. 32 — Inspecting Piston And Slippers

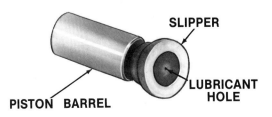

sure all shafts and other parts move freely in the pump and motor before replacing the transmission in the vehicle.

Hydrostatic transmissions can fail due to damage to any of the following components: thrust plates or fixed swashplate, piston/slipper assembly, slipper retainer, ball guides, cylinder block, bearing plates, bi-metal bearing plates, valve plates, shaft seal, charge pump assembly, displacement control valve, servo sleeve and piston, and shafts. For detailed information on each of these types of failures, refer to "Fundamentals of Service — Identification of Parts Failures."

After the transmission is reinstalled, fill it with the proper grade of new hydrostatic fluid (Fig. 27), and be sure oil has a chance to move completely through the system before activating the hydrostatic drive. (Always replace the filter whenever transmission repairs are made.)

TESTING THE SYSTEM

After a hydrostatic drive is repaired, specific tests may be recommended for charging and operating pressures, oil flow rate, and oil temperature. Use the proper test equipment and procedures as recommended in the technical service manual. Test equipment for specific systems may be available from the equipment manufacturer's service department.

TROUBLESHOOTING HYDROSTATIC DRIVES

Shown below are some of the major problems, possible causes, and normal remedies for hydrostatic drives. Note however that some items may not apply to simpler systems. For specifics, refer to the machine technical service manual.

Possible Cause	Remedy
Machine Will Not Move In Either Direction	
System low on oil	1. Check oil level.
	2. Check for oil leaks.
Plugged filter	1. Replace filter.
Faulty control linkage	1. Be sure control linkage is connected properly and not binding or broken.
Disconnected coupling	1. Be sure couplings from engine to pump and from motor to gear train are installed correctly and are not slipping or broken.
Low or zero charge pressure	1. Check pressure control valve. Also look for binding pump drive shaft or oil line restrictions.
Low or fluctuating charge pressure	1. Check for air in system (if noisy). Also look for damaged pressure control valve (stuck open) or faulty check valves.
Damaged pump or motor	1. Repair or replace pump or motor as necessary.
Machine Moves In One Direction Only	
Faulty control linkage	1. Be certain control linkage is connected properly and not binding or broken.
High pressure relief valve stuck open	1. Replace valve which is stuck open.
One faulty check valve	1. Replace both check valves.
Faulty directional control valve	1. Check speed control.
Neutral Hard To Find	
Faulty speed control linkage	1. Adjust control linkage. Be sure linkage is not binding. Also be sure displacement control valve is adjusted properly. Check pump servo valves for proper adjustment.

Possible Cause	Remedy

System Operating Hot

Overloading the system	1. Reduce load.
Oil level low	1. Fill with oil.
Cooler clogged	1. Clean cooler air passages.
Filter plugged	1. Replace filter.
Engine fan belt slipping or broken	1. Tighten or replace fan belt.
Internal leaks (usually shown by loss of acceleration and power)	1. Check for stuck high pressure relief valve. Replace relief valve. Check for internal damage in pump or motor.

System Noisy

Air in system	1. Check oil supply. Be sure all air is out of system. Check for loose fittings and damaged lines or hoses.
Incorrect type of fluid	1. Change to proper hydrostatic fluid.
Damaged pump or motor	1. Repair or replace pump or motor.

High Loss Of Oil

Loose connections or leaking lines and hoses	1. Tighten connections. Replace damaged lines and hoses. Check O-rings and seats.

Acceleration And Deceleration Sluggish

Air in system	1. Remove all air from system.
	2. Check for low oil supply.
Control orifice partially blocked	1. Check orifice and spool in displacement control unit for foreign material. If orifice and spool are clean, remove charge pump and clean passage between charge pump and displacement control unit with compressed air.
Internal damage in pump or motor	1. Repair pump or motor as necesary.

Transmission Hard To Shift Or Will Not Shift

Speed control lever not in neutral	1. Check and adjust linkage to be sure lever is in neutral.
Gearshift linkage out of adjustment	1. Adjust gearshift linkage in all positions.

CHAPTER 6 REVIEW

1. (Fill in each blank) Hydrostatic drives use _____ pressure oil at _____ speeds to transmit power.

2. What is meant by a "closed circuit" in a hydrostatic drive?

3. (Choose one word from each set) Variable-displacement pumps and motors have (fixed, movable) swashplates, but (fixed, movable) swashplates are used in fixed-displacement pumps and motors.

4. Name the three factors which control hydrostatic drive operation.

5. What team of parts is called the heart of a hydrostatic drive?

6. (True or false) A hydrostatic drive can put out more power than the engine puts into the pump.

7. (Choose one) If a pump or motor has variable displacement, and the other has fixed displacement, will the drive be in (forward, neutral, reverse) when the movable swashplate is in the vertical position?

8. What does the charge pump do in a hydrostatic drive?

9. Name the biggest enemies of hydrostatic drives and hydraulic systems.

10. What is one of the major problems related to service and repair of hydrostatic drives?

CHAPTER 7

DIFFERENTIALS

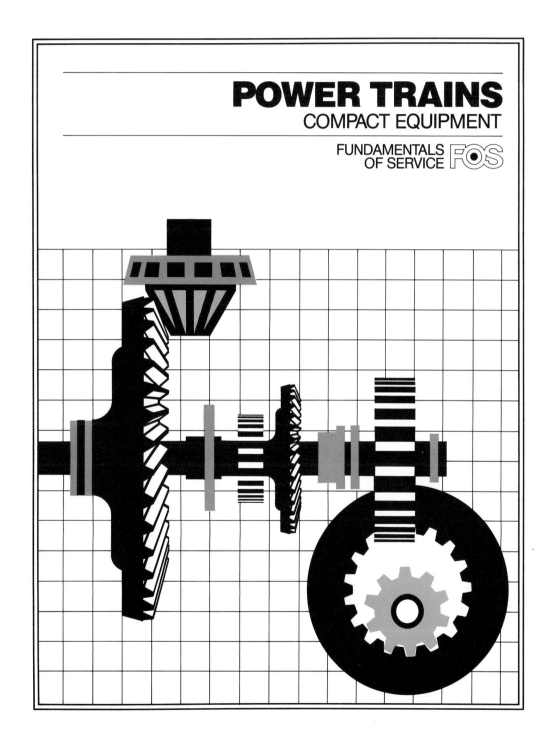

POWER TRAINS
COMPACT EQUIPMENT

FUNDAMENTALS
OF SERVICE FOS

SKILLS AND KNOWLEDGE

This chapter contains basic information that will help you gain the necessary subject knowledge required of a service technician. With application of this knowledge and hands-on practice, you should learn the following:

- **How differentials operate.**

- **How transaxles operate and where they differ from a differential.**

- **Types of differential locks used and how they operate.**

- **Maintenance and repair procedures for differentials.**

- **Troubleshooting differentials.**

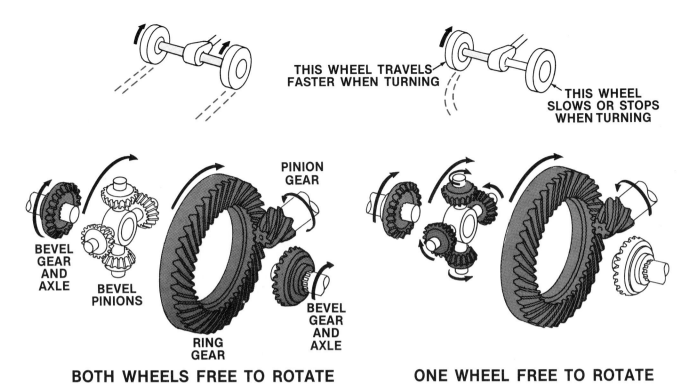

THIS WHEEL TRAVELS FASTER WHEN TURNING

THIS WHEEL SLOWS OR STOPS WHEN TURNING

PINION GEAR

BEVEL GEAR AND AXLE

BEVEL PINIONS

RING GEAR

BEVEL GEAR AND AXLE

BOTH WHEELS FREE TO ROTATE

ONE WHEEL FREE TO ROTATE

Fig. 1 — Operation Of A Differential

INTRODUCTION

Differentials perform two jobs:

● *Transmit power "around the corner" from transmission output shaft to drive axles.*

● *Permit each drive wheel to rotate at a different speed and still propel its own load.*

The ring gear and bevel gears direct power to the axles (Fig. 1), while the bevel pinions permit differential action between the ring gear and drive axles.

OPERATION OF DIFFERENTIALS

Some differentials have four bevel pinions (Fig. 1), but other simpler units (Fig. 2) have only two bevel pinions. In Fig. 2, flat surfaces in the bevel gears engage flattened sides on the right- and left-hand axles. The two bevel pinion gears mesh directly with the bevel gears on the axles and there is no ring gear and pinion. Power enters the differential through the chain sprocket shown attached to the assembled differential (Fig. 2). When the machine moves straight ahead, the bevel pinions rotate with the differential housing, and both axles turn at the same speed. As the machine turns, the outside wheel starts rotating faster and the bevel pinions "walk between" the bevel gears so the inside wheel can slow down.

In Fig. 1, engine power enters the differential through the pinion gear which drives the ring gear. When the machine is moving straight ahead, the bevel pinions and bevel gears on the drive axles all turn as a unit with the ring gear. The drive axle on each side receives equal power, so each wheel turns at the same speed.

When the machine turns a sharp corner and one wheel stops (Fig. 1), the differential permits the other wheel to rotate faster. Engine power still enters the differential on the pinion gear and drives the ring gear, but now bevel pinions are carried around with the ring gear. And, because the right-hand wheel is held stationary, the bevel pinions must rotate on their own axis and "walk around" the right-hand bevel gear.

The bevel pinions are in constant mesh with both bevel gears, but when the right bevel gear (drive axle) stops, the ring gear transmits all turning force through the bevel pinions to turn the left-hand bevel gear (drive axle).

During one revolution of the ring gear, the left-hand gear makes **two** revolutions — one with the ring gear, and another as the bevel pinions "walk around" the right-hand bevel gear.

As a result, when the drive wheels encounter unequal resistance (turning or different surface conditions), the

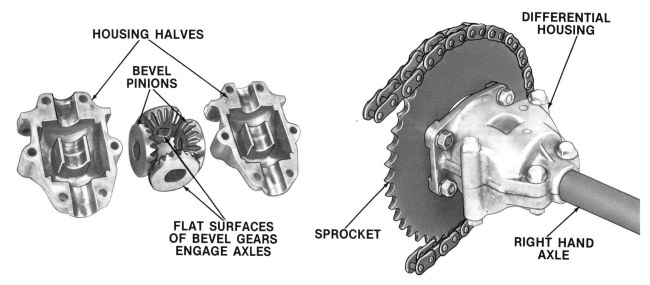

Fig. 2 — Simple Differential Without Ring Gear And Pinion

wheel with the least resistance turns more revolutions. As one wheel turns faster, the other turns slower than normal by the same amount. However, both wheels still propel their own loads, but at different speeds.

Differential action is necessary to permit smooth turns, but differential action can be a disadvantage when the wheels slip. For example, if a tractor is pushing snow and one wheel loses traction on a patch of ice, that wheel may slip or start spinning freely while the other wheel holds. The driving power is limited or falls to zero by the amount of power directed to the wheel that spins.

To prevent this power loss, differential locks or limited-slip differentials are used in many machines. These devices will be discussed later.

OPERATION OF TRANSAXLES

On some smaller machines, such as riding mowers and lawn and garden tractors, the transmission and differential are designed as a single unit and placed in the same gear case (Fig. 3). Such units are called transaxles.

Power enters the transmission (Fig. 3) through the input bevel gear which is splined or keyed to the countershaft. Two spur gears are also attached to and rotate with the countershaft.

A cluster gear splined to the shifter shaft is moved from side to side by a shifter fork.

In low range, the large gear on the cluster gear is meshed with the spur gear on the countershaft. (A smaller gear is driving a larger gear for more torque, low speed.)

Shifting to high range moves the smaller gear on the cluster gear to mesh with the larger spur gear on the countershaft. This increases output speed and reduces torque.

In high range or low range, the shifter shaft drives the output pinion which drives the differential. The differential transfers power to both axle shafts causing rear wheels to rotate. When the machine is turning, or one wheel encounters less resistance than the other, the differential permits one wheel to rotate faster, as shown in Fig. 1.

In neutral (shown in Fig. 3), the cluster gear sets between the two spur gears. The countershaft spins freely causing no movement in the rest of the transaxle.

The transaxle shown in Fig. 3 provides two speed ranges for a garden tractor with a hydrostatic transmission. Transaxles can also be used with variable speed belt drives, or be coupled directly to the engine through a belt drive (with idler tensioning clutch), or through a friction clutch and mechanical drive shaft. The transaxle may provide from two to five forward speeds and one or more speeds in reverse, depending on transaxle design and machine operating requirements.

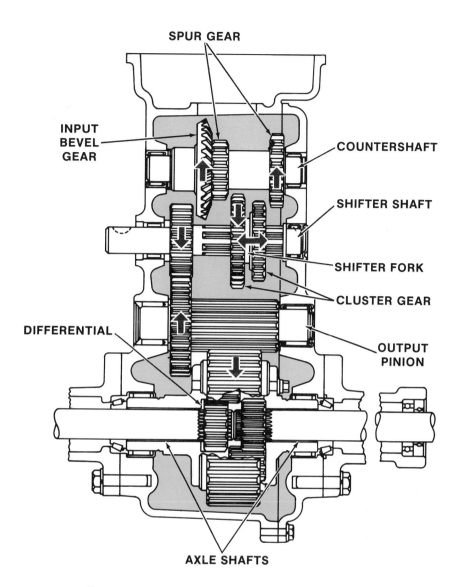

SPUR GEAR

INPUT BEVEL GEAR

COUNTERSHAFT

SHIFTER SHAFT

SHIFTER FORK

CLUSTER GEAR

DIFFERENTIAL

OUTPUT PINION

AXLE SHAFTS

Fig. 3 — Transaxle Is A Combination Transmission And Differential

In a typical transaxle (Fig. 4), power flows to the differential (Fig. 5) through a straight ring gear instead of the customary pinion and bevel gear. Inside the ring gear, the pinion shaft is forced by the drive blocks to rotate with the ring gear. As long as the vehicle moves straight ahead, power flows equally through the pinion gears to each axle shaft. However, the pinion gears are not fixed on the pinion shaft, and if the vehicle turns, or one wheel starts slipping, the pinion gears start to "walk around" the bevel gears on each axle shaft and thus permit one axle (wheel) to turn faster than the other one.

DIFFERENTIAL LOCKS

Differential locks direct power equally to both wheels by locking out the differential. This prevents the usual loss of traction when one wheel is slipping.

Some differential locks may be engaged any time one wheel starts spinning. However, others **should not** be engaged while one wheel is turning appreciably faster than the other one. The clutch must be disengaged and power flow stopped before engaging the lock. As soon as speed equalizes or wheels stop, the lock may be

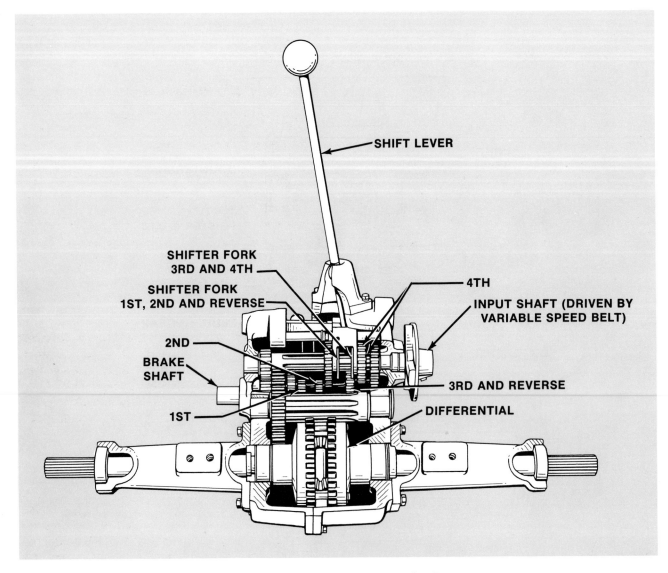

Fig. 4 — Transaxle With Four Forward Speeds And One Reverse

Fig. 5 — Transaxle Differential

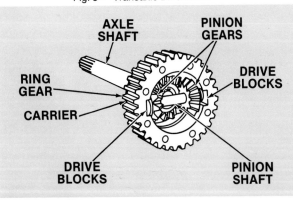

engaged and the clutch pedal released. Power then flows equally to each wheel and will usually prevent the machine from becoming stuck due to loss of traction on one wheel. The differential lock on most current machines disengages automatically as soon as traction equalizes for both drive wheels. If unequal traction is not continuous (wheels slip — gain better traction — slip again), the differential lock pedal or hand lever may be held in the engaged position to prevent automatic unlocking of the differential.

IMPORTANT: *Do not engage the differential lock when the machine is turning because one wheel must be free then to rotate faster than the other.*

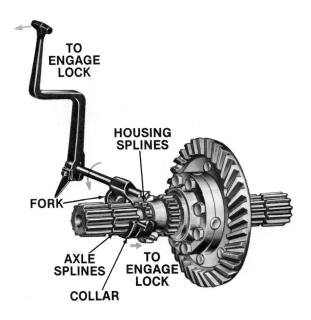

Fig. 6 — Mechanical Differential Lock On A Tractor

MECHANICAL DIFFERENTIAL LOCK

The mechanical differential lock is the type used on compact equipment.

When the operator moves the hand lever of the mechanical differential lock (Fig. 6), the fork rotates and slides the collar to the right. This engages the differential housing splines to the collar. But the collar is already splined to the axle, so both the axle and differential housing rotate as a unit. This prevents the bevel gear of the axle from turning and so locks the differential.

The collar is spring-loaded and releases automatically as soon as the torsional forces are gone. This happens when the drive wheels regain equal traction, or when power flow to the differential is stopped (the clutch is disengaged).

On some machines, the differential lock is also disengaged whenever the turning brake is applied to either drive wheel.

LIMITED SLIP DIFFERENTIALS

Manually-engaged limited slip differentials (Fig. 7) are used on some compact equipment such as walk-behind snow blowers. In the unit shown in Fig. 7, the right and left-hand axles are hollow and the axle shaft is pinned to the right axle. Flattened sides on the end of each axle engage flat surfaces in the bore of each bevel gear in the differential. Tightening the wing nut by hand or with the operator's foot flattens the spring washer and presses the limited-slip plate against the wheel hub.

The tighter the wing nut is tightened, the less differential action is permitted between the drive wheels. If the operator wishes to release the lock for easier turning, the wing nut is loosened.

⚠ CAUTION: Always stop the snow blower and disengage both the auger drive and power to the wheels before attempting to engage or disengage the limited-slip differential. If one wheel starts slipping, do not attempt to tighten the lock nut while the machine is operating.

MAINTENANCE AND REPAIR OF DIFFERENTIALS

Differentials normally operate in closed gear cases which are filled with oil. The oil must be drained at recommended intervals and replaced with the proper gear lubricant specified by the manufacturer. On some tractors, the transmission and differential cases serve as a combined reservoir for hydraulic oil which lubricates the power train and operates hydraulic equipment.

If gear lubricant is too thick because of low temperatures or failure to change oil when needed, it may not flow

Fig. 7 — Simple Limited-Slip Differential

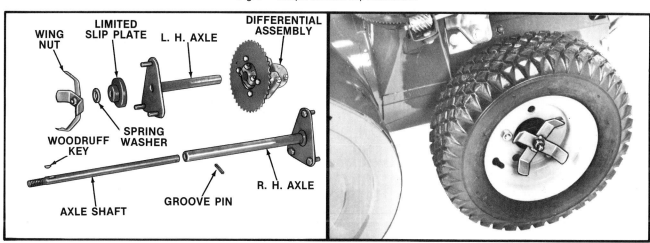

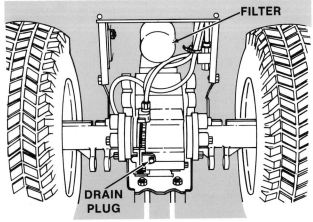
Fig. 8 — Changing Differential Oil

If the differential, transmission, and hydraulic system share a common oil supply, remove the filter also (Fig. 8). Be sure to replace the filter with a new one which meets the manufacturer's specifications. Coat the filter gasket with new oil used in the differential and tighten it only hand-tight to permit easier removal next time.

After oil has completely drained from the differential, replace the drain plug and tighten it securely. Refill the differential with the recommended grade and type of oil and follow instructions in the operator's manual or technical service manual for moving oil through the system before operating the vehicle again, particularly if differential oil supply is shared with a hydrostatic transmission.

REPAIRING THE DIFFERENTIAL

If differential repairs are required, remove the differential by following the step-by-step procedures shown in the machine technical service manual. For a typical tractor the following steps are usually needed.

● *Support the differential case on a suitable stand and remove both rear wheels.*

● *Drain the transmission/differential oil and remove the rockshaft housing from the top of the differential case.*

● *Remove both final drive housings (Fig. 9) and remove the differential lock shifter fork.*

freely around moving parts and thereby permit metal-to-metal contact between gears. This can cause rapid wear and premature differential failure.

Before changing differential oil, operate the machine long enough to bring the oil to full operating temperature. This will help ensure removal of as much old oil and contaminants as possible.

Remove the drain plug (Fig. 8) and catch old oil in a container. (Do not permit oil to run on the ground to avoid contaminating surface or ground water.) Note that on some machines you may need to remove more than one drain plug to drain all the old oil. Loosen or remove the filler plug to permit air to enter the differential as oil is drained.

Fig. 10 — Differential Housing And Ring Gear

Fig. 9 — Removing Final Drive For Access To Differential

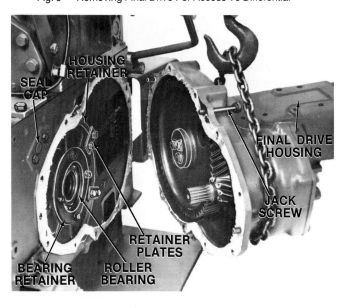

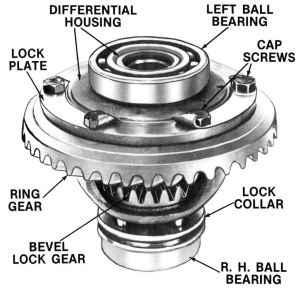

● *Remove capscrews from the retainer plate-to-transmission case on each side.*

● *Drive differential housing from the case using the proper driving tools.*

IMPORTANT: Do not drop the differential as it is removed from the case. A long drift through the differential housing will ease removal.

After inspecting the differential (Fig. 10) for wear or damage, remove only those parts which require replacement or which must be taken off for access to other parts. For example, differential housing bearings may be damaged during removal, so unless the bearing or lock collar behind must be replaced, do not remove bearings. If the differential ring gear is replaced, the differential drive shaft must also be replaced because they are a matched set and must be installed together. If the ring gear is removed and will be reinstalled, mark the ring gear and differential housing so they may be reassembled in the same position. (The ring gear must be removed on the unit shown to obtain clearance for pinion shaft removal.)

To remove the pinion gears, drive the spring pin from the differential housing and pinion shaft (Fig. 11), and remove the shaft from the housing. Next, walk the pinions around the gears and remove them through the side of the housing. Then remove thrust washers.

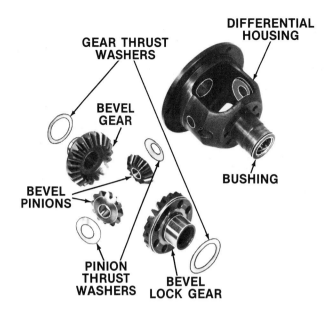

Fig. 12 — Check Parts For Wear And Damage

Inspect thrust washers, bearings, gears, and the housing (Fig. 12) for wear, cracks, or damage and replace any defective parts. Check maximum and minimum diameters of bores in gears and the housing as well as the pinion shaft and replace any parts not within the tolerances specified in the technical service manual. Measure other tolerances as instructed and replace parts as necessary.

When reassembling the differential, some parts, such as the differential ring gear and bearings, may require heating before installation. This causes those parts to expand and slip over the mating parts without requiring excessive force. As the parts cool they grip very tightly without the distortion that could result from forcing cold parts together. Always use a bearing heater to heat parts and a thermometer to avoid exceeding the recommended temperature or bearings and seals could be damaged. Also, be prepared to handle heated parts safely to avoid being burned.

Be sure any parts marked during removal are replaced in the proper relationship, and be sure recommended tolerances are provided between parts as the differential is reassembled and replaced in the machine.

Adjusting differential backlash places the differential ring gear in proper relation to the spiral bevel gear from side to side. Always check backlash if ring gear and bevel pinion were replaced, or if there is question concerning the proper number of side shims to be used.

Fig. 11 — Removing Pinion Shaft

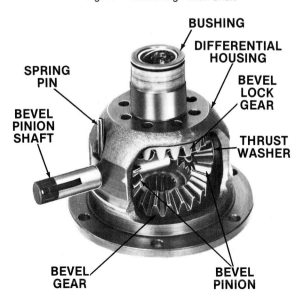

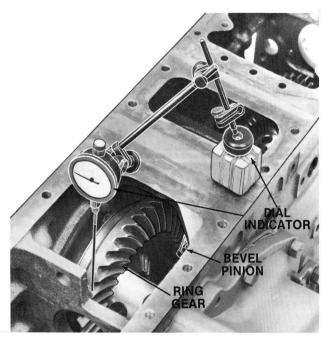

Fig. 13 — Checking Differential Backlash

Carefully follow the manufacturer's procedure for checking backlash so that all parts are in the proper relationship. For example, in a typical tractor, the dial indicator is installed on the case (Fig. 13) and the differential lock engaged to force the differential housing and left-hand bearing against the bearing retainer. The differential drive shaft is turned several revolutions to ensure that all parts are properly positioned.

After releasing the differential lock, the bevel pinion is held stationary and the differential ring gear rotated and the dial indicator movement noted. Recommended backlash for this unit is 0.005-0.007 inch (0.13-0.18 mm). If backlash does not fall within the prescribed range, side shims must be added or removed and backlash rechecked.

Remember that the design of each differential is different, even within the same product line. Consequently, there are different points to look for whenever you disassemble and repair various units.

Then, in addition to looking for obviously damaged or broken parts, inspect the entire differential for wear or damage. In a typical differential (Fig. 14), check particularly for these points:

Fig. 14 — Exploded View Of Typical Differential Assembly

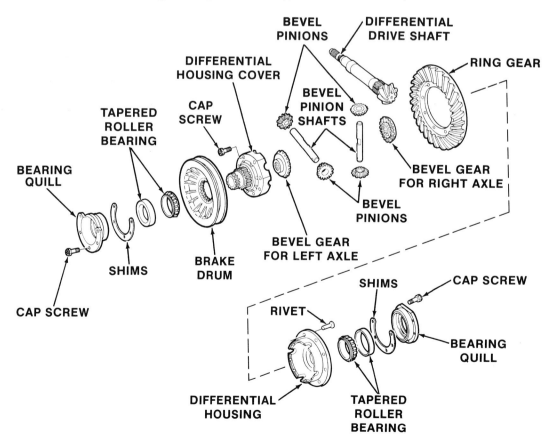

1. Examine the thrust and bearing surfaces of bevel gears on each axle. If bevel gears are worn or damaged, bores in the differential housing and cover may also be damaged.

2. If the differential housing must be replaced, separate the housing from the ring gear by drilling out the rivets.

3. Rivet the new differential housing to ring gear by heating rivets and alternately installing rivets on opposite sides of housing. Avoid delays when riveting. Heated rivets cool quickly and cannot be riveted correctly.

4. Check bevel pinions and bevel pinion shafts for excessive wear. If these parts are damaged or badly worn, also replace parts they mesh with or parts they slide on.

5. Check ring gear for wear and damage. If ring gear must be replaced, replace the complete ring gear and differential housing assembly.

IMPORTANT: If the ring gear and differential housing assembly is replaced, also replace the differential drive shaft. These parts are commonly furnished as matched sets and are not available individually for replacement.

6. Check roller bearings for wear and serviceability.

7. Check bearing quills (carriers) for wear and damage.

8. Coat bevel gears, pinion gears, and pinion shafts with clean new transmission oil before reinstalling them in the differential.

9. Install differential housing cover on differential housing and tighten screws to the recommended torque.

If the differential failed during operation or was badly worn, drain all old oil from the case and flush it with diesel fuel or solvent to remove any remaining metal particles which could cause rapid wear of replaced parts.

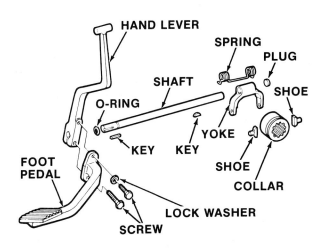

Fig. 15 — Exploded View of Differential Lock

After the differential is reinstalled in the case, use a dial indicator to check endplay and backlash of the differential assembly. Refer to the technical service manual for the proper endplay or backlash. If necessary, add more shims (Fig. 14) until the correct movement is obtained. When all parts are assembled and adjusted, fill the case with the correct amount of the recommended gear lubricant or oil **before** starting the engine.

If the differential lock is removed for repairs:

1. Examine the shifter fork and shoe guides for wear and damage (Fig. 15).

2. Check splines of shifter and of the differential housing cover for wear or damage.

3. Replace springs and keys as necessary.

4. Check the shaft for wear or distortion.

5. Check linkage and replace parts as necessary.

TROUBLESHOOTING DIFFERENTIALS

Look for these common differential failures and the possible causes:

Trouble	Possible Cause
Noise in differential at all times	1. Incorrect adjustment of ring gear and pinion.
	2. Worn or damaged ring gear or pinion.
	3. Damaged bearings on pinion shaft.
	4. Damaged bearings in differential housing.
	5. Low oil level in differential case.
	6. Improper oil used in differential case.
Differential not working freely on turns	1. Damaged or galled bearing surfaces between bevel gears or between bevel pinions and differential housing.
	2. Damage or galling between bevel pinions and their shafts.
	3. Low oil level in differential case.
	4. Improper oil used in differential case.
Mechanical differential lock does not hold	1. Linkage between engaging lever and shifting collar broken or improperly adjusted.
	2. Shifting collar broken or damaged.
Limited-slip differential will not hold	1. Wing nut on axle not tightened enough.
	2. Limited-slip plate and spring washer badly worn or saturated with oil.
	3. Pin or key sheared in axle or limited-slip clutch.

CHAPTER 7 REVIEW

1. What two jobs does the differential do?

2. (True or false) All differentials have four pinion gears between the ring gear and the axle bevel gears.

3. (Choose the correct answer) When a differential permits wheels to change speeds in a turn, one wheel slows by (the same, twice) the amount that the other one speeds up.

4. What is a transaxle?

5. What is the purpose of a differential lock?

6. Can the differential lock be engaged while the machine is moving?

CHAPTER 8

FINAL DRIVES

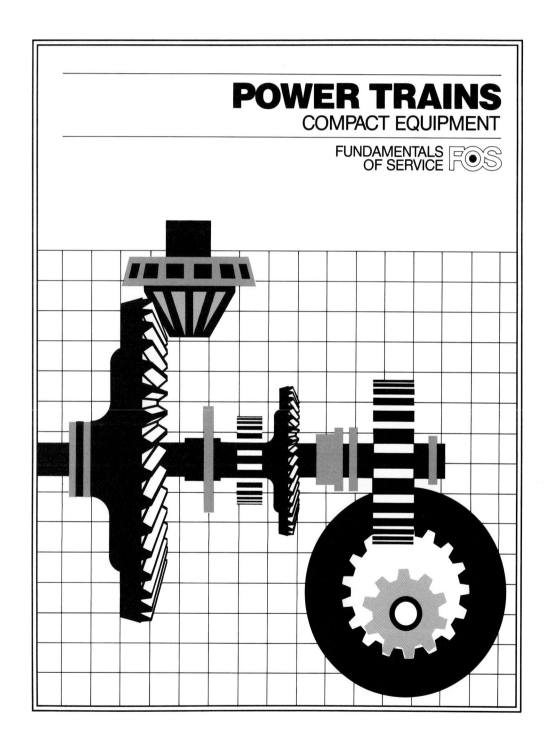

POWER TRAINS
COMPACT EQUIPMENT

FUNDAMENTALS
OF SERVICE FOS

SKILLS AND KNOWLEDGE

This chapter contains basic information that will help you gain the necessary subject knowledge required of a service technician. With application of this knowledge and hands-on practice, you should learn the following:

- **Types of final drives used.**

- **Operating principles of planetary drives.**

- **Operating principles of chain drives.**

- **Operating principles of worm gear drives.**

- **Maintenance and repair procedures for final drives.**

- **Causes of final drive failures.**

- **Procedures for adjusting final drives and axle bearings.**

- **Troubleshooting procedures for final drives.**

Fig. 1 — Final Drive In A Modern Tractor

INTRODUCTION

Power is transferred from the differential to machine drive wheels through the final drive (Fig. 1). The final drive also provides the final reduction in speed and increases torque to the drive wheels.

Final drives are mounted near the drive wheels of most wheeled machines as shown in the four-wheel drive configuration in Fig. 2. However, on equipment such as walk-behind tillers, which have no drive wheels, the final drive carries power to the tiller tines.

By reducing speed in the final drive, stress is reduced in the transmission and other power train components and lighter weight gears and shafts can be used. However, on some machines the final drive must support the weight of the machine as well as withstand torque and shock loads during operation.

TYPES OF FINAL DRIVES

Five types of final drives are commonly used in compact equipment:

- **Straight axle**
- **Pinion**
- **Planetary**
- **Chain**
- **Worm gear**

Each of these drives has specific advantages for different applications, and all but the first can provide a speed reduction.

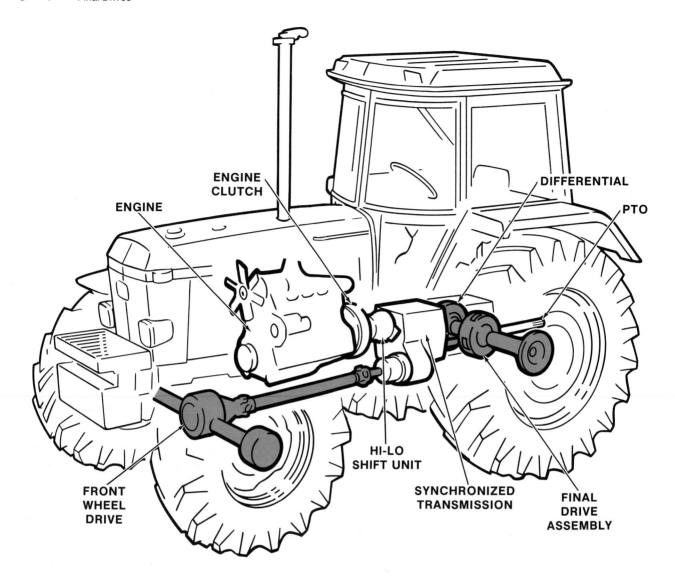

Fig. 2 — Final Drives In A Front Wheel Drive Tractor

STRAIGHT AXLE DRIVES

Straight axle drives are simple and easy to maintain and repair. Straight axle final drives are commonly used in automobiles, light-duty trucks, lawn and garden tractors, riding mowers, and similar equipment where extra gear reduction is not needed for heavy loads.

The drive wheels in a straight axle receive their power from the differential. For each revolution of the differential, the axle shaft and wheels make one revolution.

Machine axles may be referred to as "live" axles which transmit engine torque to drive wheels and in most cases support part of the machine weight; or "dead" — those which only support machine weight, such as unpowered tractor front axles.

Two types of straight axle construction are used:

- **Rigid axle shaft**

- **Flexible axle shaft**

The **rigid axle shaft** (Fig. 3) is commonly connected to the differential output by a splined coupling and may or may not be enclosed in an axle housing. If an axle housing is used, it usually forms an integral part of the differential case, making one long housing from drive wheel to drive wheel.

The **flexible axle shaft** is used for independently suspended drive wheels and is sometimes called a "swing axle." The axles are connected to the differential by universal joints and permit each wheel to move up or down

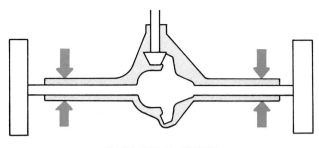

RIGID AXLE SHAFT

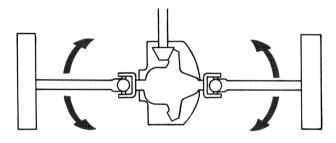

FLEXIBLE AXLE SHAFT

Fig. 3 — Two Types Of Straight Axle Drives

without affecting the position of the differential or transmission. Flexible axle shafts are often used in compact automobiles (especially those with engine in the rear), but are not common in compact equipment.

Semifloating Axle Shafts In Rigid Drives

Semifloating axle shafts are supported at one end by the differential output and at the other end by a single bearing assembly positioned between the axle shaft and housing or frame member (Fig. 4). Therefore, the shaft supports a portion of the vehicle weight as well as transmitting engine torque to the wheels. The axle shafts must absorb stresses of turning, skidding, etc.

Either tapered roller bearings or ball bearings are used in semifloating axles. When tapered roller bearings are used, an adjustment for axle shaft endplay is provided. This is done with shims or an adjusting nut at one end of the axle housing. To transmit shaft end thrust equally to the bearings at both ends of the axle housing, a thrust block or spacer is located between the inner ends of the two axle shafts in the differential.

Semifloating axles are commonly used in many tractors, automobiles, light trucks, and compact equipment. The outer ends of the axle shafts may be flanged or tapered as shown in Fig. 4. When the shaft is tapered, the wheel hub is keyed to the shaft and locked in place with a nut.

PINION DRIVES

A spur gear and pinion are used for the final gear reduction on many tractors and other machines. This enables the differential to transmit power at high speeds and lower torque.

The type of pinion and gear reduction used most commonly is located within the differential case.

Advantages of having the pinion gear reduction within the differential are:

● *All gears are enclosed within the differential-transmission case. This is a more compact unit and only one lubrication system is needed.*

● *The final drive is in a straight line, permitting use of a long, straight axle shaft. This can provide a wide range of wheel tread adjustments for different applications.*

Power is transmitted to the drive wheels through pinion gears which are connected to the differential output. The pinions mesh with the larger final drive gears which in turn rotate the axle shafts (Fig. 5).

Fig. 4 — Two Types Of Semifloating Axle Shafts

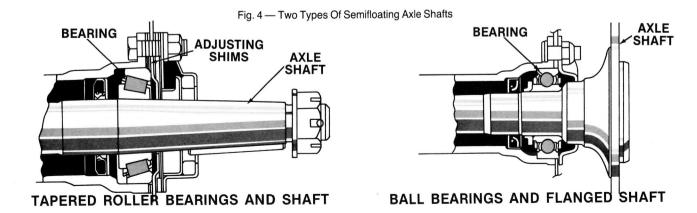

TAPERED ROLLER BEARINGS AND SHAFT   **BALL BEARINGS AND FLANGED SHAFT**

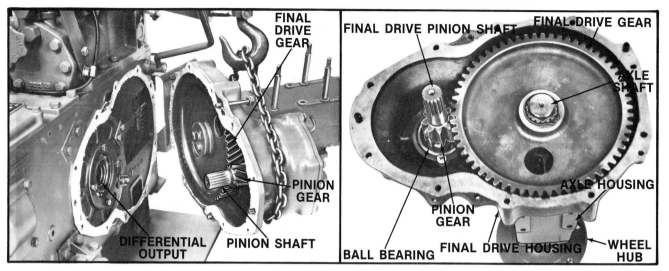

Fig. 5 — Pinion Gear Reduction Within Differential Case

A final drive gear is splined to the inner end of each axle shaft, and each shaft is supported by two bearings — one in each end of the axle housings. Bearing adjustment is made by adding or removing shims from the axle shaft. However, some similar drives are equipped with an adjusting nut to control endplay.

The axle shafts support vehicle weight and absorb end thrust as well as transmitting engine torque to the wheels.

PLANETARY DRIVES

The planetary gear reduction is smaller and more compact than the pinion type. Planetary systems are also more durable under heavy loads since torque loads are spread more evenly over several gears.

Planetary gear final drive systems are common on farm and utility tractors, industrial equipment, and trucks.

Fig. 6 — Final Drive With Planetary Gears Next To Differential (Inboard Type)

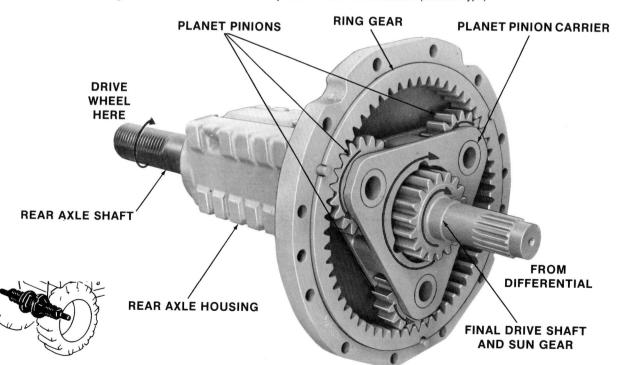

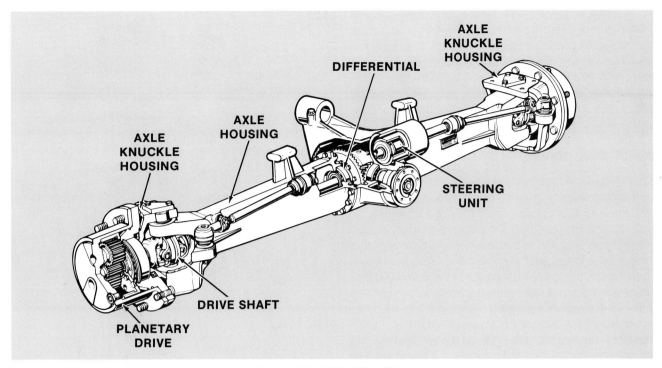

Fig. 7 — Front-Wheel Final Drive

The planetary gear set may be next to the differential (Fig. 6) or at the outer ends of the final drive (Fig. 7). Both types — called inboard and outboard — operate the same.

The planetary gear reduction system receives engine power from the differential through the final drive shaft and sun gear. The sun gear is an integral part of the final drive shaft (axle shaft) and so turns with the shaft. The sun gear meshes with the planet pinions which are mounted in the planet pinion carrier.

As the sun gear turns it forces the planet pinions to "walk around" the inside of the ring gear as shown by the arrows. The planet pinion carrier is forced to rotate in the same direction as the sun gear, thus delivering engine power to the drive wheels **but at a reduced speed and an increased torque.**

Construction of Planetary Next To Differential

With this planetary (Fig. 6), a straight in-line axle can be used where a range of wheel tread adjustments is needed much as on farm tractors. In addition, all gears are located in one compact housing and use a common oil supply for lubrication.

The axle shaft is carried on bearings in each end of the axle housing, and the inner end of the shaft is splined to the planet carrier (Fig. 6).

With this type of axle, the axle shaft supports vehicle weight and absorbs end thrust, as well as transmitting engine torque.

Construction of Planetary at Outer Ends of Final Drives

When planetary gear systems are located at the outer ends of the final drive, the drive system is usually enclosed in one rigid housing (Fig. 7). This solid unit includes the differential housing, both axle housings, and the planetary gear housings.

Fig. 8 — Front-Wheel Drive Showing Drive Shaft

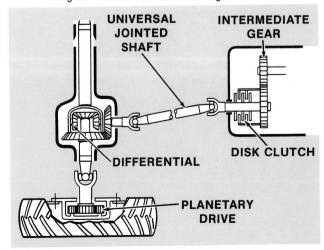

On tractors equipped with front wheel drive, engine power is forwarded from the transmission via a disk clutch (Fig. 8) to a universal-jointed shaft. This shaft forwards power to differential of the front axle. Here the power is forwarded at right-angles through drive shafts to planetary drives in each wheel hub. The planetary drives give a reduction in speed and at the same time, an increase in torque.

CHAIN FINAL DRIVES

Chain final drives are used in many compact skid-steer loaders, mowers with zero turning radius, and similar machines to direct power independently to wheels on each side of the machine. On some machines, engine power flows through a hydrostatic pump to separate motors on each side of the machine. From the hydrostatic motor, chains deliver power to each drive wheel. On other machines, separate hydrostatic pumps and motors are used for each side.

The skid-steer loader in Fig. 9 uses friction clutches and multiple chains to drive each side of the machine. The upper jackshaft (Fig. 9) is belt-driven from the engine. Forward and reverse clutches are controlled by a single lever (not shown) which engages one clutch as the other clutch is disengaged. Note that the lower jackshaft chain passes under sprocket "R" and over sprocket "F." This permits reversing the rotation of the lower jackshaft by engaging one clutch and disengaging the other. A neutral position in the center disengages both clutches, and power flow to wheels stops on that side of the machine.

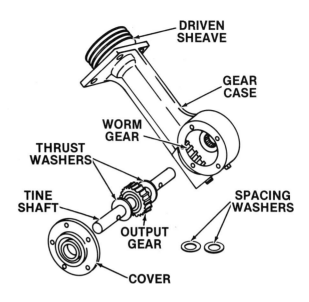

Fig. 10 — Worm Gear Final Drive

Engaging the forward clutch on one side and leaving the other drive in neutral causes the machine to turn. Engaging the forward clutch on one side and the reverse clutch on the other side causes the machine to pivot within its own tracks.

So, in this type machine, a chain final drive is used, but there is no differential. (Another chain drive with forward and reverse clutches like the one shown in Fig. 9 is used to control the other side of the machine.) Differential action is provided by the operator who controls power flow to each side of the machine by engaging and disengaging the clutches. This permits maneuvering the machine in any manner desired — forward, reverse, wide turns, tight turns, etc.

WORM GEAR FINAL DRIVE

In a worm gear drive (Fig. 10), a V-belt delivers engine power to the input shaft. (A dual-belt drive may be used to provide forward and reverse drives if desired. See Dual-Belt Drives, Chapter 5, Transmissions.) The worm gear changes the direction of power flow and drives components such as rotary tiller tines attached to the tine shaft.

A worm gear drive is smooth-running, quiet, and more durable than a chain drive. The output speed in relation to input speed depends on the respective diameters of the worm gear and output gear. A large worm gear and small driven gear would provide faster tine speed compared to a smaller worm gear and larger output gear. A 21 to 1 reduction is provided by the gears shown.

Running the worm gear in oil lubricates the drive and cools moving parts.

Fig. 9 — Chain Final Drive For One Side Of Skid-Steer Loader

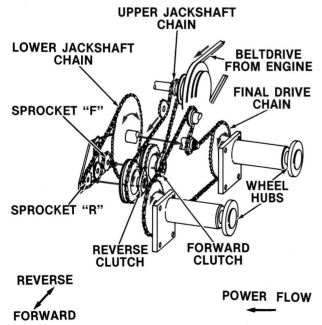

Fig. 11 — Checking Final Drive Lubricant

MAINTENANCE AND REPAIR OF FINAL DRIVES

The reliability of any final drive depends on good maintenance, operating at rated load, and proper repair if a failure occurs.

Most final drives require little or no maintenance. The only maintenance usually necessary is to check the oil level once a year in the differential (Fig. 11). For satisfactory long-term operation, final drives must be provided with the proper lubricant (usually SAE 90 oil) as recommended by the manufacturer. If you can feel the oil when the plug is removed, the oil level is all right. Chain tension must also be adjusted whenever needed.

Watch key points when diagnosing final drive failures:

- **Excessive shaft endplay**

- **Overheating**

- **Lack of lubrication**

EXCESSIVE SHAFT ENDPLAY

Excessive shaft endplay is normally caused by loose bearings. However, it can also be created by:

- *Foreign material in the lubricant which causes rapid bearing wear.*

- *Overloading the machine with either weight or engine torque.*

- *Poorly adjusted bearings at time of assembly.*

A continuous noise or knock in the final drive is a strong indication of loose or damaged bearings. On machines with semifloating axles, the noise can be heard in the differential as the ends of the axle shafts rap the spacer block. Readjustment of worn or damaged bearings will not provide a satisfactory repair. Bearings must be replaced.

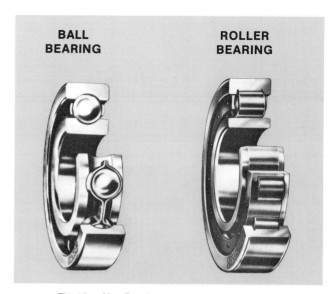

Fig. 12 — New Bearings Are Shiny And Turn Freely

Compare the roller end of a new tapered bearing with a worn bearing. If worn rollers have no shoulder compared to the new bearing, replace the old bearing.

New bearings (Fig. 12) appear smooth and shiny, turn freely on their races, and have very little freeplay between the races and balls or rollers. However, worn bearings generally have much more movement between the races. And, as you hold the inner race of a used bearing, rotate the outer race and feel for any drag, roughness, or scratching. Severely worn bearings (Fig. 13) often appear scratched or pitted by rust and dirt caused by poor lubrication or dirt and water entering the bearing. Such bearings must be replaced.

Fig. 13 — Damaged Bearings Must Be Replaced

SCRATCHES   RUST

Fig. 14 — Removing An Axle Bearing

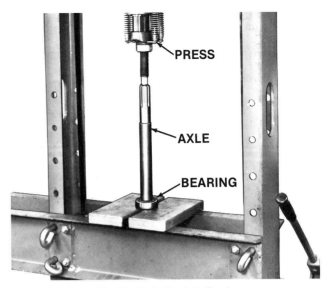

Fig. 15 —Installing Axle Bearing

To remove prefit bearings, support the inner bearing race on a press bed and apply pressure to the splined end of the shaft (Fig. 14). Do not use extremely hard drivers in the press to avoid damaging or distorting the end of the axle. Also, be sure the bearing is supported squarely on the press bed and that the axle is positioned directly under the press shaft.

To replace a press-fit bearing, support the inner bearing race on the press bed and insert the axle shaft through the bearing as far as possible (Fig. 15). Be sure to turn the bearing seal in or out as specified in the technical service manual. Carefully align the bearing, axle shaft, and driver so that the axle is pressed squarely into the bearing. The end of the axle could be distorted or damaged if pressure is applied off-center or if excessive pressure and a very hard driver are used. Be sure the bearing is properly seated, but attempting to force bearing too far onto the axle could distort the bearing race.

OVERHEATING

Many final drives are damaged simply by overheating. This is usually caused by not maintaining the lubricant at the proper level or using the wrong type of lubricant.

Overloading or abusing the machine can also cause overheating. Excessive loads cause deflection of the final drive assembly and concentrate stresses which accelerate wear.

In addition, improper assembly, excessive preload, or inadequate endplay in shafts can also increase friction in the final drive and result in overheating.

Galling, pitting, or scoring on the surface of mating parts indicates lack of lubricating film and that overheating has occurred. Extreme overheating may result in bluish or burned areas on gears, shafts, and bearings.

LACK OF LUBRICATION

Loss of lubricant through worn or broken oil seals and gaskets or loose drain plugs can severely damage the final drive. While some bearings are automatically lubricated by oil in the gear case, or oil creeping along the drive shaft from the differential, others are sealed off and may require separate lubrication. Refer to the operator's manual or technical service manual for instructions on lubricating the final drive. Avoid excessive greasing of sealed bearings as this can damage the seals and permit entry of dirt and water to the bearing.

Before installing new oil seals, be sure they are flexible and pliable where they fit around the shaft. Soak or coat the seal in clean new oil used in the final drive before installing it. Avoid scratching or damaging the seal during installation.

Fig. 16 — Removing Final Drive (Axle) Housing

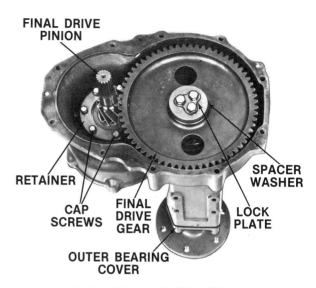

Fig. 17 — Disassembly Of Final Drive

DISASSEMBLY AND ASSEMBLY OF FINAL DRIVES

For access to the final drive, follow these steps:

• Remove fenders, rollover protective structures (ROPS), and other attachments to the axle housings.

• Support the rear of the tractor so there is no danger of tipping while repairs are made.

• Remove rear wheels.

• Support axle housing with a chain hoist and disconnect it from differential case (Fig. 16).

• Remove capscrews and washers holding final drive gear (Fig. 17) and remove gear.

• Drive axle shaft from the housing. If outer bearing cover is left on the housing, the outer bearing and seal will be removed during axle removal (on the unit shown — others may be different). Avoid damaging the end of the axle as it is removed.

• Remove other parts as necessary for inspection or replacement.

When the final drive is disassembled, inspect all parts and provide replacements as necessary. Check for:

• Excessively worn or chipped pinion teeth and splines.

• Worn, chipped, or damaged gears.

• Leaking or damaged oil seals and noisy, loose, or damaged bearings.

• Shaft diameters worn beyond minimum tolerance for bearings.

When reassembling the final drive, follow the step-by-step procedures outlined in the technical service manual. Also, be sure to use proper tools such as seal drivers (Fig. 18) to press seals and bearings evenly into place.

FINAL DRIVE ADJUSTMENTS

To prevent premature failures, adjust the final drive properly. This includes adjusting the axle bearings — either preloading or permitting a specified endplay.

Fig. 18 — Use Seal Drivers And Other Special Tools For Assembly

ADJUSTING AXLE BEARINGS

Loose gears, or bearings that are too tight or too loose, will break down prematurely, regardless of their strength and design.

Bearings must be adjusted to the manufacturer's specifications. Some designs require the bearings to be preloaded while others require a slight amount of shaft endplay. Do not make the mistake of permitting endplay in a final drive built to have a bearing preload, or vice versa. Follow instructions in the technical service manual.

PRELOADING BEARINGS

When the manufacturer's instructions call for preloading bearings, bearings must be placed under slight tension. In some cases this is done by **SLIGHT** overtightening of the bearing adjusting nut. In other designs it is done with shims. Preload is normally measured in inch-pounds (Newton-meters) of rolling or rotating torque (**NOT** starting torque).

Some torque specifications will include frictional drag such as oil seals and gears in the final drive. If it is not included, measure the rolling torque while the axle shaft has a slight amount of endplay, and **add** this torque reading to the specified torque. Then make the necessary bearing preload adjustment. Also, when new bearings are installed, a greater amount of torque will be required. Always make the preload adjustment in small increments until the proper preload is obtained.

Two methods are generally used to measure rolling torque.

One method is to use a special tool or adapter with a torque wrench (Fig. 19). This gives a direct reading on the wrench.

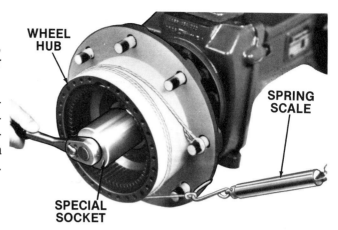

Fig. 20 — Checking Bearing Preload With Pull Scale

Another method of checking preload is with a cord and pull scale (Fig. 20). The preload is figured by multiplying the radius (distance from the center of the drive shaft or axle to a point on the circumference from which the cord is pulled) by the reading in pounds on the scale.

For example, assume the radius is 4 inches (100 mm) and the scale reading is 7 pounds (31 N). Multiply 4 inches (100) by 7 pounds (31) and you have 28 inch-pounds (3.1 N·m) of rolling torque (100 x 31 = 3100 N·mm = 3.1 N·m).

INSPECTING A FINAL DRIVE

Very careful inspection of final drive parts while the drive is disassembled and replacement of worn or damaged parts as needed can eliminate the need for having to do the job over again too soon. Check the following points:

● Inspect teeth on pinion or planetary gears. Replace gears if teeth are chipped, damaged, or worn excessively.

● Measure diameter of shafts at bearings and be sure shafts are within specified tolerances. Replace worn or damaged shafts.

● Inspect shaft and gear splines for wear or damage and replace as necessary.

● Replace any oil seals which have been leaking or that are worn or damaged. Some manufacturers suggest replacing **all** oil seals removed during repair work.

● Inspect all bearings and replace any bearings with damaged seals, bearings which fit loosely on their shafts, or those which are noisy when rotated. On some machines, **all** pinion shaft needle bearings are replaced whenever the final drive is repaired.

● Replace damaged or bent retaining or snap rings.

● Inspect all locking pins, keys, and keyways for wear or damage and replace parts as necessary.

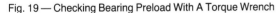

Fig. 19 — Checking Bearing Preload With A Torque Wrench

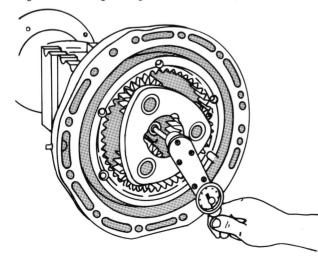

TROUBLESHOOTING FINAL DRIVES

Unless the machine is severely or frequently overloaded, final drives are generally quite troublefree. However, recognizing the types of problems which can be encountered and what causes these problems will reduce repair time and should help avoid similar problems in the future.

Trouble	Possible Cause
Loss of lubricant	1. Worn or damaged oil seals or gaskets
	2. Loose or missing drain plugs
	3. Cracked or damaged housing
	4. Too much or improper lubricant
Unusual noise when running	1. Pinion or planetary gears worn or damaged
	2. Worn planet pinion bearing rollers
	3. Worn or damaged bearings
	4. Insufficient or improper type of lubricant used
Overheating	1. Insufficient or improper type of lubricant used
	2. Overloading or abusing the machine
	3. Improper assembly, inadequate endplay, or excessive preload

CHAPTER 8 REVIEW

1. Name five types of final drives.

2. (True of false) A semifloating axle shaft carries no vehicle weight.

3. List two advantages of locating a pinion-type final drive within the differential case.

4. List two advantages of planetary gear reductions compared to pinion-type drives.

5. What is the main disadvantage of a chain final drive?

6. (True or false) Worm gears cannot be used to change speed in a final drive.

7. List three common causes of final drive failures.

8. (True or false) A final drive designed for a specified bearing preload should have no shaft endplay.

CHAPTER 9

POWER TAKE-OFFS

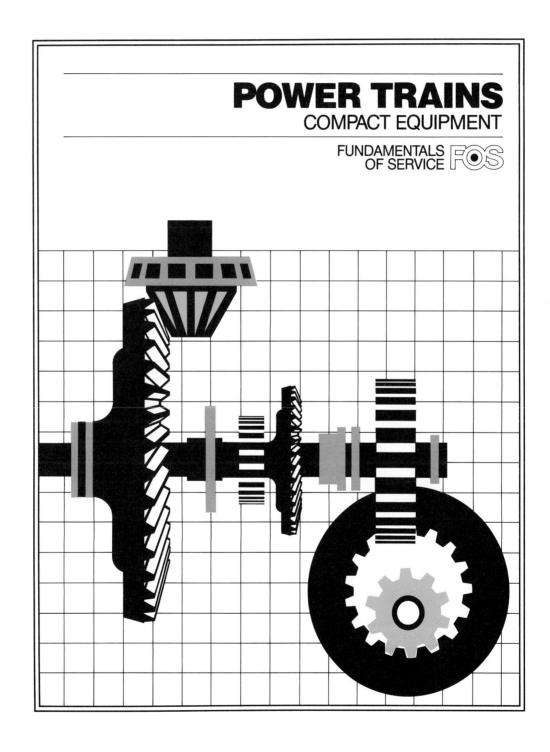

POWER TRAINS
COMPACT EQUIPMENT

FUNDAMENTALS
OF SERVICE FOS

SKILLS AND KNOWLEDGE

This chapter contains basic information that will help you gain the necessary subject knowledge required of a service technician. With application of this knowledge and hands-on practice, you should learn the following:

- **Types of power take-offs used.**

- **Operation, similarities, and differences between solid and telescoping PTO shafts.**

- **Operation of transmission-driven PTO controls.**

- **Operation of continuing-running PTO controls.**

- **Operation of independent PTO controls.**

- **How a power take-off operates.**

- **PTO shielding standards used.**

- **Safety rules for operating PTO shafts.**

- **Maintenance and repair of PTO shafts.**

- **Troubleshooting PTO shafts.**

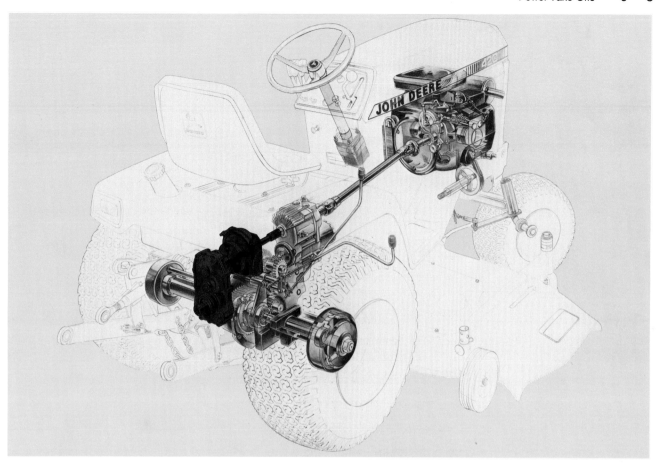

Fig. 1 — Power Take-Off (PTO) For Modern Tractor

INTRODUCTION

Power take-off (PTO) shafts (Fig. 1) deliver engine power to auxiliary equipment or working components which may be mounted on the vehicle (tractor, mower, etc.) or trailed behind.

In this chapter we will look at power take-off shafts on the basis of:

- **Types of PTO shafts.**
- **Types of PTO control.**
- **Standards for PTO drives.**
- **Safety rules for PTO drives.**
- **PTO service and maintenance.**

TYPES OF POWER TAKE-OFFS

Power take-off is defined as a removable, flexible, splined shaft (and related parts) used to transmit power from the machine to some other equipment.

However, power to drive auxiliary equipment may also be "taken-off" through belt drives or hydraulic outlets. Belt drives are particularly common on smaller equipment such as lawn and garden tractors, and we have discussed their service and maintenance in Chapter 2. As we have pointed out before, a detailed discussion of hydraulic systems is beyond the scope of this book. Therefore, this discussion will focus on mechanical drives within the PTO definition given above.

Universal Joints

To compensate for misalignment and permit relative movement between the power outlet and PTO-driven equipment, universal joints (U-joints) are provided at each end of the PTO shaft.

Three types of U-joints are commonly used, depending on the amount of power transmitted and other drive requirements. These joints are:

- **Cardon or Hooke joints.**
- **Bendix-Weiss joints.**
- **Flexible disk joints.**

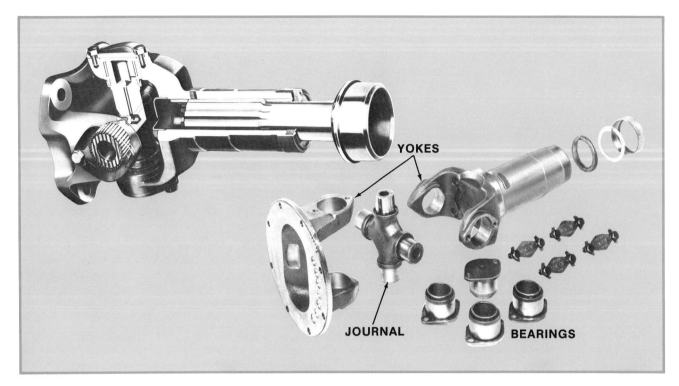

Fig. 2 — Cardon or Hooke Universal Joint

The most common type is the Cardon or **Hooke joint** (Fig. 2). This joint consists of two U-shaped yokes fastened to the ends of the two shafts to be connected. Inside these yokes is a cross-shaped journal which holds the yokes together and permits each yoke to bend or pivot with respect to the other. This means the drive will operate even when one shaft is as much as 30 degrees out of alignment with the other.

However, any time the Hooke joint operates at an angle, speed of the driven shaft fluctuates. It still takes the same time for the shaft to make one complete revolution. But the yoke design causes the shaft to accelerate and decelerate twice in each revolution. These speed fluctuations increase as the angle between the drive shaft and the driven shaft increases. But in a PTO drive with two U-joints, proper shaft assembly cancels out these speed changes and the implement speed remains constant.

To avoid fluctuations in the speed of machine operation when the PTO shaft operates at an angle, make certain

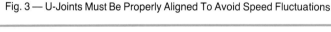

Fig. 3 — U-Joints Must Be Properly Aligned To Avoid Speed Fluctuations

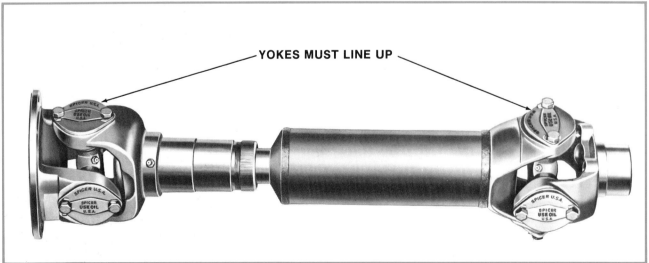

the center acts as a spacer to hold the larger balls in place.

The Bendix-Weiss joint transmits rotary motion at a constant rate, but it is more expensive and has less strength than the Hooke joint. Consequently, the Hooke joint is more commonly used on farm and industrial equipment and compact equipment.

Flexible disk joints (Fig. 5) are used for economical, low-power applications in some lawn and garden equipment. The drive yoke and shaft yoke are each bolted to the reinforced plastic disks at 90 degrees to each other. This permits the disks to flex and compensate for minor shaft misalignment between the engine and driven component. However, flexible disks are not designed to flex enough to operate equipment which is raised and lowered by the tractor or for turning corners with trailed equipment.

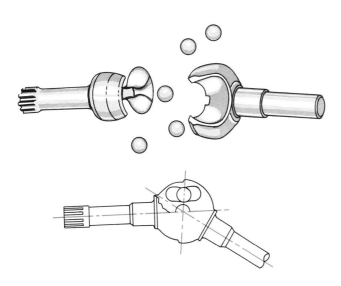

Fig. 4 — Bendix-Weiss U-Joint

that the yokes on each end of the center shaft section are in line (Fig. 3). Incorrect assembly is usually prevented by minor differences in shaft dimensions and positioning of keyways or splines. However, some older, square telescoping PTO shafts can be assembled wrong and thus multiply the speed changes and pass those changes to the driven equipment any time the shaft is out of alignment. This places unnecessary stress on the entire power train and must be avoided.

The **Bendix-Weiss** U-joint (Fig. 4) provides smoother power flow when operating out of alignment, but is limited to low-power applications. Four large balls are used to transmit rotary force through the joint; a smaller ball in

PTO SHAFTS

Two types of PTO shafts are commonly used.

- **Solid shafts.**
- **Telescoping shafts.**

SOLID PTO SHAFTS

The solid PTO shaft has a fixed length and a U-joint at each end. The distance between the two joints is always the same. Therefore, this type of drive is used only where there is no relative movement between the power outlet and the driven equipment, but where it is necessary to compensate for possible misalignment between the power outlet and the driven equipment. Flexible disk joints commonly use solid PTO shafts.

Fig. 5 — Flexible Disk Coupling — Parts And As Attached To Engine

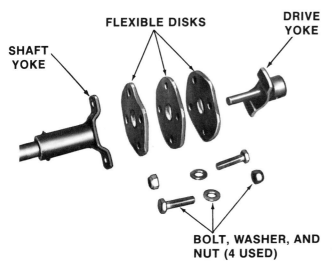

FLEXIBLE DISKS

DRIVE YOKE

SHAFT YOKE

BOLT, WASHER, AND NUT (4 USED)

FLEXIBLE DISK COUPLING

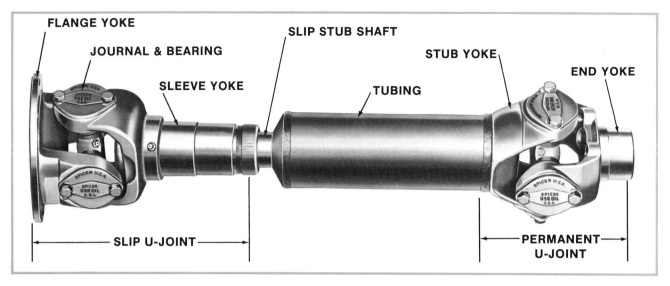

Fig. 6 — Telescoping PTO Shaft

TELESCOPING PTO SHAFTS

Telescoping PTO shafts have a sliding element and usually have two U-joints. They may, however, have three U-joints in some applications.

Telescoping shafts with two U-joints (Fig. 6) have one joint at the drive end and another at the equipment (driven) end of the shaft. A slip joint between these two points permits the shaft to telescope in and out as the PTO-driven equipment moves in relation to the power outlet on the machine.

The telescoping PTO shafts commonly used on compact equipment have a square or rectangular shaft which slides inside a matching steel tube. The mating parts must be kept well lubricated to permit them to slide freely. However, some telescoping shafts on large machines have ball bearings in the telescoping members to permit easier movement between the mating parts during operation.

In a PTO drive with three U-joints, a telescoping PTO shaft is connected to the power outlet by a U-joint. The

Fig. 7 — Three-Joint PTO Shaft In Working And Transport Positions

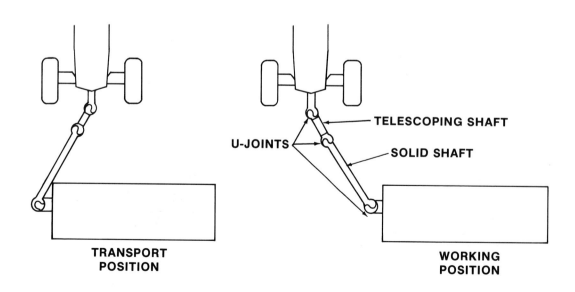

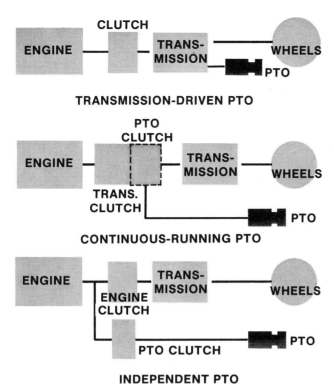

CONTINUOUS-RUNNING PTO

INDEPENDENT PTO

Fig. 8 — Three Types Of Power Take-Off (PTO) Control

U-joint at the other end of the telescoping shaft is attached to the forward end of a solid PTO shaft supported by a bearing mounted on the frame or hitch of the equipment (Fig. 7). The other end of the solid shaft connects to the equipment drive through the third U-joint. This provides the flexibility of a telescoping two-joint shaft, plus the ability to change hitch angle without disconnecting the drive when equipment is converted to transport position.

TYPES OF POWER TAKE-OFF CONTROL

The three basic types of PTO control are:

- **Transmission-drive.**
- **Continuous-running.**
- **Independent.**

The three types are compared in Fig. 8.

Transmission-driven PTO shafts operate only when the engine clutch is engaged. And they stop whenever the clutch is disengaged (stopping machine travel). However, on most machines the PTO can be engaged and operated when the transmission is in neutral.

Continuous-running PTO shafts have two clutches — one for the PTO and one for the transmission. Both clutches are operated by the same control. Depressing

the clutch pedal halfway (Fig. 9) operates the transmission clutch while the second half of the pedal travel operates the PTO clutch. This permits stopping tractor travel without interrupting operation of PTO-driven equipment.

For momentary stops, such as to avoid plugging of PTO-driven equipment, the clutch can be held halfway down to disengage power flow to the machine drive wheels until the equipment clears. Then the pedal can be released and normal operation resumed. However, if the transmission is shifted to neutral while the PTO is operating, the PTO clutch must be disengaged too before the transmission can be shifted into gear again.

Independent PTO shafts have a separate clutch completely independent of the engine clutch and transmission. This means that the PTO can be engaged or disengaged while the machine is stopped or in motion (not possible with transmission-driven and continuous-running PTO shafts). It also means that machine travel can be started and stopped and gears shifted without affecting PTO operation (also not possible with other types of PTO control).

OPERATION OF PTO

Most PTO shafts are gear-driven from the transmission or directly from the engine and send power through a shaft to the PTO outlet where the driven equipment is coupled.

The PTO output shaft on some garden tractors is belt-driven from the engine and controlled by either an electric clutch in the driven sheave or by a belt tensioning clutch.

Fig. 9 — Two-Stage Clutch Controls Continuous-Running PTO

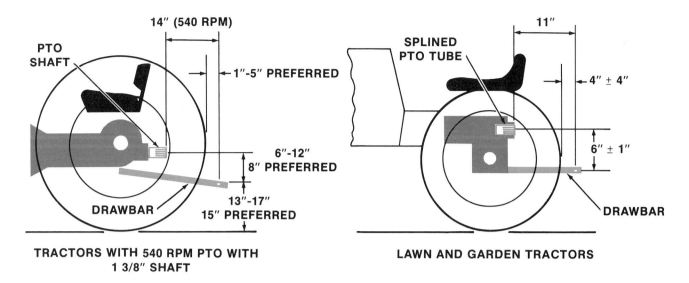

Fig. 10 — Tractor Dimensions Related To PTO Standards

STANDARDS FOR PTO DRIVES

The American Society of Agricultural Engineers (ASAE) and the Society of Automotive Engineers (SAE) have worked together for years to establish standards for tractor and implement features. These standards improve safety and help when equipment is interchanged from one tractor to another.

TRACTOR PTO STANDARDS

Standards have been developed for two different PTO types in the compact tractor class:

Some tractor dimensions related to PTO operation (Fig. 10) have been standardized according to the PTO used on that tractor. This permits easier attachment and operation of equipment made by different manufacturers.

These standards prescribe preferred, and in some cases maximum and minimum, dimensions for such features as height of the drawbar from the ground; height of the PTO shaft above the drawbar; and distance from the end of the PTO shaft to drawbar hitch pin.

In addition, the lateral location of the PTO shaft should be within 1 inch (25.4 mm) of the tractor centerline. The

Type	Speed (Rpm)	Shaft Diameter (Inches)	Number Of teeth On Shaft
1	540	1-3/8	6
L&G*	2000	1	15

*Tentative lawn and garden tractor PTO standard.

NOTE: Tractors meeting this standard have a splined tube into which a splined shaft is inserted rather than a stub shaft as on larger tractors.

Note that all ASAE and SAE standards are "voluntary" standards. Manufacturers cannot be legally obligated to meet the standard requirements. Consequently, nonstandard PTO speeds and PTO-related dimensions may be found on some older agricultural tractors and on some tractors made outside the United States. The standard for lawn and garden tractors is tentative and is not yet in wide use.

Fig. 11 — PTO Shaft Guard And Master Shield

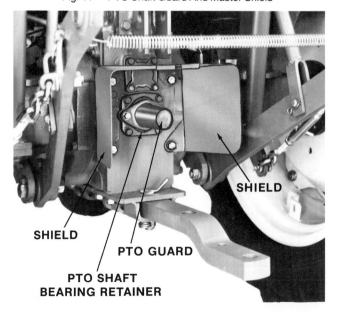

hitch point should be directly under the extended centerline of the PTO shaft and provisions made to lock the hitch in that position.

All standard PTO shafts rotate clockwise when viewed from the rear of the tractor. And standard dimensions are provided for the size and shape of the splined PTO shaft or tube used in each of the PTO types described earlier.

PTO SHIELDING STANDARDS

The tractor PTO shaft and the equipment drive shaft must be covered at all times to prevent accidental injury to operators or bystanders. Unshielded rotating shafts can catch clothing or body parts and cause serious injury or death.

The tractor PTO shaft guard and master shield (Fig. 11) are provided by the tractor manufacturer, while the implement shaft shielding is provided by the equipment maker.

The ASAE and SAE standards provide basic dimensions for the tractor PTO master shield which permit convenient coupling and uncoupling of the drive shaft to the machine. They also list requirements for shielding the drive shaft to the equipment.

Two basic types of equipment drive line shields are used:

- **The tunnel shield**
- **The spinner shield**

Tunnel shields are inverted, U-shaped steel plates which cover the top and sides of the moving parts. They must be coupled and uncoupled each time equipment is attached or detached from the machine. For this reason,

some operators are tempted to leave them off. This is a serious mistake which could lead to injury or death.

The spinner shield has largely replaced tunnel shields because it offers more complete protection from moving parts and is simpler to use. The spinner shield (Fig. 12) is a steel cylinder around the PTO shaft with a bell housing on each end to cover the universal joints. It does not have to be coupled and uncoupled to attach and detach equipment.

The spinner shield is free to rotate with the shaft. If you accidentally press your leg against it, the spinner shield stops rotating and keeps your leg from contacting the shaft spinning freely inside the shield. Ball bearings between the spinner shield and the drive shaft permit the shield to stop quickly at any time. However, a bent spinner shield or bearings that stick can prevent the shield from stopping when it should. This is quite dangerous, and damaged shields and bearings should be replaced immediately.

SAFETY RULES FOR PTO DRIVES

Safety is a prime factor in PTO equipment design, operation, and maintenance. Here are several safety rules to keep in mind.

- Keep PTO guards and shields in place when the PTO is not operating.

- Always disconnect the PTO when not in use.

- Never engage the PTO while the engine is shut off. (Some PTO clutches automatically disengage when the engine stops.)

- Keep yourself and your clothing away from PTO parts.

- Always wear relatively tight, belted clothing when operating a PTO. Loose clothing can become caught in moving PTO and equipment parts.

- Never operate PTO shafts at extreme angles.

- Never ride, or permit others to ride, on the drawbar of the tractor — whether or not the PTO is being used.

- Make sure PTO rpm and equipment rpm are the same.

- Be sure PTO spinner shields rotate freely at all times. Disengage all power and shut off the engine before checking spinner shields.

- Always be sure PTO drive shaft is properly secured to the PTO shaft on the machine.

- Do not service, adjust, lubricate, or perform other work on a machine without first disengaging the machine PTO and shutting off the engine.

Fig. 12 — Spinner Shield And Lubrication for PTO Drive Shaft

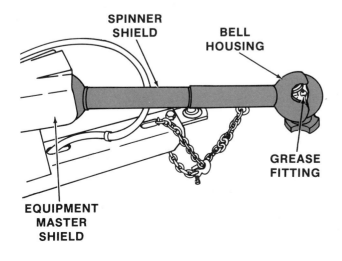

SPINNER SHIELD

BELL HOUSING

GREASE FITTING

EQUIPMENT MASTER SHIELD

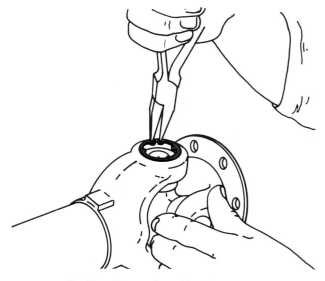

Fig. 13 — Remove Snap Rings From Yokes

MAINTENANCE AND REPAIR OF PTO DRIVES

The service life of a PTO drive depends on the maintenance the drive receives. Some drives require more maintenance than others, so always follow directions in the operator's manual and technical service manual.

At recommended intervals, lubricate (Fig. 12):

• *The telescoping drive shafts and shields.*

• *Bearings supporting the shaft and spinner shield (if greasable bearings are used).*

• *Universal joints.*

• *Other shafts and bearings on the machine PTO drive.*

At regular intervals, check clutch pedal or control lever free travel and adjust as necessary according to instructions in the manual. If the drive slips during operation, adjust or replace the clutch as necessary and avoid overloading the drive.

If the drive is noisy or vibrates when operating, check for worn or damaged bearings in the tractor PTO drive or shaft. Refer to the tractor operator's manual.

REPAIRING PTO SHAFT

Repairing universal joints is relatively simple, but care must be taken to avoid damaging parts during disassembly or assembly. To disassemble a universal joint, remove the PTO shaft from the implement or vehicle and proceed as follows:

• Remove snap rings from each yoke (Fig. 13).

• Support the drive shaft on an open vise (Fig. 14) and use a hammer and brass drift to drive the yoke downward thus exposing the bearing sleeve.

• Clamp the exposed bearing sleeve in the vise and tap the yoke from the bearing. (Avoid pounding directly on bearing sleeves, especially with a hard tool or punch which might damage sleeves and snap ring grooves in the yokes.)

• Follow the same procedure to remove the other bearing sleeve and remove the yoke.

• Remove bearing sleeves from the yoke on the drive shaft (Fig. 15) and take out the journal. Repeat this process to disassemble the joint on the other end of the shaft.

Fig. 14 — Removing Needle Bearing Sleeves

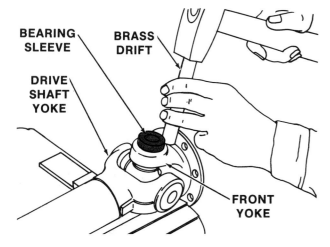

BEARING SLEEVE
BRASS DRIFT
DRIVE SHAFT YOKE
FRONT YOKE

Fig. 15 — Removing Bearing Sleeve From Drive Shaft

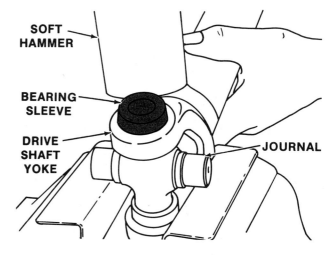

SOFT HAMMER
BEARING SLEEVE
DRIVE SHAFT YOKE
JOURNAL

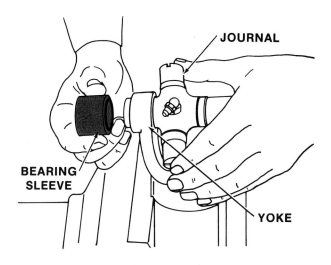

Fig. 16 — Installing Journal And Bearing Sleeve

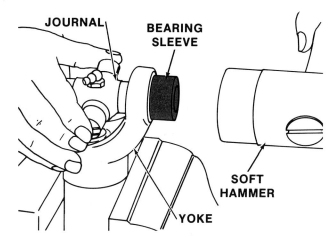

Fig. 17 — Driving Bearing Sleeve Into Yoke

After the universal joints have been disassembled, inspect all parts as follows:

● Check tubular shafts for cracks or signs of having been bent or twisted under heavy load.

● Look for cracked welds attaching yokes to drive shaft. Replace the drive shaft if it is damaged.

● Inspect journals, oil seal rings, needle bearing rollers, and bearing sleeves for wear or damage. Replace parts as needed.

● Inspect snap rings and replace if cracked or badly bent.

● Replace the complete drive shaft and universal joints if parts are badly worn or damaged.

To reassemble the universal joint:

● Apply grease to needle rollers and be sure all rollers are in place in the sleeve.

● Slip the journal through the yoke as far as possible (Fig. 16) and insert it into the sleeve to hold the needle rollers in place as the sleeve is installed.

● Use a soft-faced hammer (Fig. 17) and drive the bearing sleeve deep enough into the yoke to permit installation of the snap ring.

● Assemble other bearing sleeves and journals in the same manner.

● Thoroughly lubricate universal joints after completing assembly.

Fig. 18 — U-Joint Journal Failures

TROUBLESHOOTING PTO DRIVES

Several PTO drive problems and their possible causes are shown below. Refer to the technical service manual for repair procedures.

Trouble	Possible Causes
Excessive machine vibration and/or noise during operation	1. U-joints not properly aligned
	2. Operating drive shaft at extreme angles
	3. Worn or damaged bearings in machine or shaft
Twisted PTO shaft	1. Overload on PTO shaft
	2. Improper hitch setting
PTO shaft not telescoping properly	1. Rusty, bent, or damaged spinner or tunnel shield
	2. Telescoping shaft members bent or damaged
	3. Telescoping shaft or shield members not lubricated properly
	4. Worn bearings
	5. Overloading drive shaft
Brinnelling of U-joint journal (Fig. 18)	1. Excessive metal fatigue
	2. Poor heat treatment of parts
	3. Inadequate clearance between parts (improper assembly)
Galling of U-joint journal cross ends (Fig. 18)	1. Drive shaft rpm too high
	2. Inadequate lubrication
	3. Operating drive shaft at extreme angles
	4. Inadequate clearance between parts (improper assembly)
Journal cross ends and cups chipping	1. Overloading U-joint capacity
	2. Inadequate lubrication
	3. Operating drive shaft at extreme angles
Abrasive corrosion on PTO shaft	1. Lack of lubrication
	2. Extreme low angle operation

CHAPTER 9 REVIEW

1. List two purposes of universal joints in a PTO drive.

2. Name three types of universal joints.

3. Name two types of commonly used PTO shafts.

4. Name three types of PTO control.

5. (Choose the correct answer.) Transmission gears can be shifted without interrupting PTO operation only with (transmission-driven, continuous-running, independent) PTO shafts.

6. What are the two standard PTO operating speeds?

7. Name two types of PTO drive shaft shields.

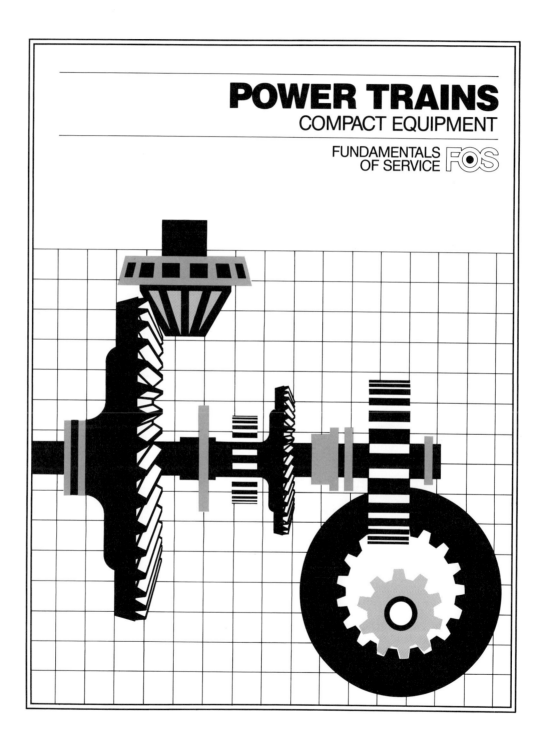

POWER TRAINS
COMPACT EQUIPMENT

FUNDAMENTALS
OF SERVICE FOS

WALK-BEHIND MOWER BLADE-BRAKE CLUTCHES

To comply with safety requirements established by the U.S. Consumer Products Safety Commission, many manufacturers have installed blade-brake clutches (BBC's) on the blades of walk-behind rotary lawn mowers. Some of these clutches fit in categories described in Chapter 4 on page 4-3 while others are combination mechanical and centrifugal devices. All are mounted between the mower engine and the cutting blade and are required to stop the blade within three seconds after the operator releases the control handle. The control must be held continuously in the engaged position to keep the blade turning. Four general types of blade-brake clutches are being used.

CONE CLUTCH

This clutch "locks" the blade to the engine when the operator engages the clutch control on the mower handle. When the control is released, the clutch is disengaged and four brake blocks move outward against the brake housing. The brake blocks are spring-loaded and self-energizing to stop the blade quickly.

FRICTION DISK CLUTCH

A single friction disk is moved to contact different surfaces for clutching action when the control is engaged and blade braking when the control is released. Springs move the disk in the appropriate direction for engagement or disengagement. A larger contact area is provided for braking than for clutching so that the brake will outlast the clutch, thus increasing safety of the system.

MECHANICAL-CENTRIFUGAL CLUTCH

This hybrid mechanical-centrifugal clutch has a built-in flywheel pinned directly to the engine crankshaft. An expanding-shoe clutch is attached to the blade. When the control is engaged, the brake is released and clutch shoes move outward against the flywheel assembly to drive the blade at direct engine speed. Releasing the control causes the brake band to pull clutch shoes away from the flywheel and engage the brake to stop the blade.

CENTRIFUGAL CLUTCH

Engagement of this clutch is tied to engine speed. As speed increases, the clutch engages, then disengages automatically when engine or blade speed drops off, for instance in thick grass. This helps prevent engine stalling under heavy load. Fly weights engage the brake disk for quick blade stoppage when the control handle is released.

MAINTENANCE AND REPAIR

All of these blade-brake clutches are relatively new, and each requires its own specific maintenance and repair procedures. Brake-clutch units should be kept clean and linkage must be lubricated and adjusted when necessary as recommended by the manufacturer.

Repairs and adjustments of these units must follow the manufacturer's prescribed methods to maintain proper function and safe stopping requirements. In fact, special tools are required to disassemble some units to discourage do-it-yourself user repairs. Some brake-clutch units are designed for replacement of individual parts by qualified dealer service technicians. For other units, manufacturers specify return or replacement of subassemblies or the complete unit. Because of these requirements and limitations, you must consult and follow the technical service manual for repair of each mower model. For these reasons no service instructions are provided for mower blade-brake clutches in this book.

DEFINITIONS OF TERMS

A

ABRASION—Wearing or rubbing away of a part.

ALIGNMENT—An adjustment to a line or to bring into a line.

ANTIFRICTION BEARING—A bearing constructed with balls, rollers, or the like between the journal and the bearing surface to provide rolling instead of sliding friction.

ASAE—American Society of Agricultural Engineers.

ASME—American Society of Mechanical Engineers.

AXIAL—Parallel to the shaft or bearing bore.

AXLE—The shaft or shafts of a machine upon which the wheels are mounted.

B

BACKLASH—The clearance or "play" between two parts, such as meshed gears.

BALL BEARING—An antifriction bearing consisting of a hardened inner and outer race with hardened steel balls which roll between the two races.

BDC—Bottom dead center.

BEARING—The supporting part which reduces friction between a stationary part and a rotating part.

BEVEL SPUR GEAR—Gear that has teeth with a straight center-line cut on a cone.

BONDED LINING—A method of cementing brake linings to shoes or bands which eliminates the necessity of rivets.

BRAZE—To join two pieces of metal with the use of a comparatively high melting point material. An example is to join two pieces of steel by using brass or bronze as a solder.

BREAK-IN—The process of wearing-in to a desirable fit between the surfaces of two new or reconditioned parts.

BROACH—To finish the surface of metal by pushing or pulling a multiple-edge cutting tool over or through it.

BURNISH—To smooth or polish by the use of a sliding tool under pressure.

BUSHING—A removable liner for a bearing.

C

CARRIER—An object that bears, cradles, moves, or transports some other object or objects.

CASE-HARDEN—To harden the surface of steel.

CASTELLATE—Formed to resemble a castle battlement, as in a castellated nut.

CENTER OF GRAVITY—The point at which a mass of matter balances. For example, the center of gravity of a wheel is the center of the wheel hub.

CENTRIFUGAL FORCE—A force which tends to move a body away from its center of rotation. Example: whirling a weight attached to a string.

CHAMFER—A bevel or taper at the edge of a hole or a gear tooth.

CHASE—To straighten up or repair damaged threads.

CHILLED IRON—Cast iron on which the surface has been hardened.

CLEARANCE—The space allowed between two parts, such as between a journal and a bearing.

CLOCKWISE ROTATION—Rotating the same direction as hands on a clock.

CLUTCH—A device for connecting and disconnecting the engine from the transmission or for a similar purpose in other units.

COEFFICIENT OF FRICTION—The ratio of the force resisting motion between two surfaces in contact to the force holding the two surfaces in contact.

COMPOUND—A mixture of two or more ingredients.

CONCENTRIC—Two or more circles having a common center.

CONSTANT MESH TRANSMISSION—A transmission in which the gears are engaged at all times, and shifts are made by sliding collars, clutches, or other means to connect the gears to the output shaft.

CONTRACTION—A reduction in mass or dimension; the opposite of expansion.

CORRODE—To eat away gradually as if by gnawing, especially by chemical action, such as rust.

COUNTERBORE—To enlarge a hole to a given depth.

COUNTERCLOCKWISE ROTATION—Rotating the opposite direction of the hands on a clock.

COUNTERSHAFT—An intermediate shaft which receives motion from a main shaft and transmits it to a working part. Sometimes called a lay shaft.

COUPLING—A connecting means for transferring movement from one part to another; may be mechanical, hydraulic, or electrical.

D

DEAD AXLE—An axle that only supports the machine and does not transmit power.

DEFLECTION—Bending or movement away from normal due to loading.

DENSITY—Compactness; relative mass of matter in a given volume.

DIAGNOSIS—A systematic study of a machine or machine parts to determine the cause of improper performance or failure.

DIAL INDICATOR—A type of measuring instrument where the readings are indicated on a dial rather than on a thimble as on a micrometer.

DIFFERENTIAL GEAR—The gear system which permits one drive wheel to turn faster than the other.

DIRECT DRIVE—Direct engagement between the engine and driveshaft where the engine crankshaft and the driveshaft turn at the same rpm.

DISTORTION—A warpage or change in form from the original shape.

DOUBLE REDUCTION AXLE—A drive axle construction in which two sets of reduction gears are used for extreme reduction of the gear ratio.

DOWEL PIN—A pin inserted in matching holes in two parts to maintain those parts in fixed relation one to another.

DRIVE LINE—The universal joints, drive shaft, and other parts connecting the transmission with the driving axles.

DROP FORGING—A piece of steel shaped between dies while hot.

DUAL REDUCTION AXLE—A drive axle construction with two sets of pinions and gears, either of which can be used.

E

ECCENTRIC—One circle within another circle wherein both circles do not have the same center. An example of this is a cam on a camshaft.

ENDPLAY—The amount of axial or end-to-end movement in a shaft due to clearance in the bearings.

F

FEELER GAUGE—A metal strip or blade finished accurately with regard to thickness used for measuring the clearance between two parts; such gauges ordinarily come in a set of different blades graduated in thickness by increments of 0.001 inch.

FIT—The contact between two machined surfaces.

FLANGE—A projecting rim or collar on an object for keeping it in place.

FLUID DRIVE—A drive in which there is no mechanical connection between the input and output shafts, and power is transmitted by moving oil. (See Chapter 5 on Hydrostatic Transmissions.)

FOOT POUND (or lbs. ft)—A measure of the amount of energy or work required to lift 1 pound a distance of 1 foot.

FREE-WHEELING CLUTCH—A mechanical device which will engage the driving member to impart motion to a driven member in one direction but not the other. Also known as "overrunning clutch."

G

GEAR—A cylinder- or cone-shaped part having teeth on one surface which mate with and engage the teeth of another part which is not concentric with it.

GEAR RATIO—The ratio of the number of teeth on the larger gear to the number of teeth on the smaller gear.

GRIND—To finish or polish a surface by means of an abrasive wheel.

H

HEAT TREATMENT—Heating, followed by fast cooling, to provide a required hardness or metal structure.

HEEL—The outside or larger half of the gear tooth.

HELICAL—Shapes like a coil spring or a screw thread.

HELICAL GEAR—Gears with the teeth cut at an angle to the axis of the gear.

HUB—The central part of a wheel or gear.

HYDRAULIC PRESSURE—Pressure exerted through the medium of a liquid.

HYPOID GEAR—A gear that is similar in appearance to spiral bevel gear, but the teeth are cut so that the gears match in a position where the shaft centerlines do not meet.

I

ID—Inside diameter.

IMPELLER—The pump or driving member.

INPUT SHAFT—The shaft carrying the driving gear by which the power is applied, as to the transmission.

INTEGRAL—The whole made up of parts.

J

JOURNAL—A bearing within which a shaft operates.

K

KEY—A small block inserted between the shaft and hub to prevent circumferential movement.

KEYWAY—A groove or slot cut to permit the insertion of a key.

KNURL—To indent or roughen a finished surface.

L

LAPPING—The process of fitting one surface to another by rubbing them together with an abrasive material between the two surfaces.

LINKAGE—Any series of rods, yokes, and levers used to transmit motion from one unit to another.

LIVE AXLE—The shaft through which the power travels from the drive axle gears to the driving wheels.

LOST MOTION—Motion between a driving part and a driven part which does not cause actuation of the driven part. Also see Backlash.

LOW SPEED—The gearing which produces the highest torque and lowest speed of the wheels at a given engine speed.

M

MISALIGNMENT—When bearings are not on the same centerline within good functional or working limits.

MULTIPLE DISK—A clutch having a number of driving and driven disks as compared to a single plate clutch.

N

NEEDLE BEARING—An antifriction bearing using a great number of long, small-diameter rollers; also known as a quill-type bearing.

O

OD—Outside diameter.

OSCILLATE—To swing back and forth like a pendulum.

OUTPUT SHAFT—The shaft or gear which delivers the power from a device, such as a transmission.

OVERDRIVE—Any arrangement of gearing which produces more revolutions of the driven shaft than of the driving shaft.

OVERRUN COUPLING—A free-wheeling device to permit rotation in one direction but not in the other.

P

PEEN—To stretch or clinch over by pounding with the rounded end of a hammer.

PINION—The smaller of two meshing gears.

PINION CARRIER—The mounting or bracket which retains the bearings supporting a pinion shaft.

PLANET CARRIER—The carrier or bracket in a planetary system which contains the shafts upon which the pinions or planet gears turn.

PLANET GEARS—The gears in a planetary gear set which connect the sun gear to the ring gear.

PLANETARY GEAR SET—A system of gearing which is modeled after the solar system. A pinion is surrounded by an internal ring gear and planet gears are in mesh between the ring gear and pinion around which all revolve.

PRELOAD—A load within the bearing, either purposely built in, or resulting from adjustment.

PRESS FIT—Mounting with interference, i.e., bore of bearing is smaller than OD of shaft, or OD of bearing is larger than bore of housing, or both.

PUMP—A device which produces motion in a liquid.

R

RACE—A channel in the inner or outer ring of an antifriction bearing in which the balls or rollers roll.

RADIAL—Perpendicular to the shaft or bearing bore.

RADIAL CLEARANCE (Radial displacement)—Clearance within the bearing and between balls and races perpendicular to the shaft.

RADIAL LOAD—A force perpendicular to the axis of rotation.

RATIO—The relation or proportion that one number bears to another.

REACTOR—See "Stator."

REAM—To finish a hole accurately with a rotating fluted tool.

RECIPROCATING—A back-and-forth movement, such as the action of a piston in a cylinder.

RING GEAR—A gear which surrounds or rings the sun and planet gears in a planetary system. Also the name given to the spiral bevel gear in a differential.

RIVET—A headed pin used for uniting two or more pieces by passing the shank through a hole in each piece and securing it by forming a head on the opposite end.

ROLLER BEARING—An inner and outer race upon which hardened steel rollers operate.

RPM—Revolutions per minute.

S

SAE—Society of Automotive Engineers.

SCORE—A scratch, ridge, or groove marring a finished surface.

SEAT—A surface, usually machined, upon which another part rests or seats; for example, the surface upon which a valve face rests.

SEPARATORS—A component in an antifriction bearing which keeps the rolling components apart.

SHIM—Thin sheets used as spacers between two parts, such as the two halves of a journal bearing.

SHRINK-FIT—Where the shaft or part is slightly larger than the hole in which it is to be inserted. The outer part is heated above its normal operating temperature or the inner part chilled below its normal operating temperature or both and assembled in this condition; upon cooling an exceptionally tight fit is obtained.

SLIDING-FIT—Where sufficient clearance has been allowed between the shaft and journal to allow free-running without overheating.

SLIDING GEAR TRANSMISSION—A transmission in which gears are moved on their shafts to change gear ratios.

SLIP-IN BEARING—A liner made to extremely accurate measurements which can be used for replacement purposes without additional fitting.

SPIRAL BEVEL GEAR—A ring gear and pinion wherein the mating teeth are curved and placed at an angle with the pinion shaft.

SPIRAL GEAR—A gear with teeth cut according to a mathematical curve on a cone. Spiral bevel gears that are not parallel have center lines that intersect.

SPLINE—Multiple keys in the general form of internal and external gear teeth used to prevent relative rotation of cylindrically fitted parts.

SPUR GEAR—Gears cut on a cylinder. The teeth are straight and parallel to the axis.

SQ. FT.—Square feet.

SQ. IN.—Square inch.

STATOR—The third member (in addition to turbine and pump) which changes direction fluid under certain operating conditions.

STRESS—The force to which a material, mechanism, or component is subjected.

SUN GEAR—The central gear in a planetary gear system around which the rest of the gears rotate.

T

TAP—To cut threads in a hole with a tapered, fluted, threaded tool.

TEMPER—To change the physical characteristics of a metal by applying heat.

TENSION—Effort which elongates or "stretches" a material.

THRUST LOAD—A load which pushes or reacts through the bearing in a direction parallel to the shaft.

TOLERANCE—A permissible variation between the two extremes of a specification or dimension.

TORQUE—A twisting motion, usually measured in ft-lbs.

TORQUE WRENCH—A special wrench with a built-in indicator to measure the applied force.

TRANSAXLE—Type of construction in which the transmission and differential are combined in one unit.

TRANSMISSION—An assembly of gears or other elements which gives variations in speed or direction between the input and output shafts.

TROUBLESHOOTING—A process of diagnosing the source of the trouble or troubles through observation and testing.

TUNE-UP—A process of accurate and careful adjustments to obtain the best performance.

TURBINE—A rotary device for obtaining mechanical power from a pressurized flow of gases or liquids.

TURBULENCE—A disturbed or irregular motion of fluids or gases.

V

VANES—Any plate, blade, or the like attached to an axis and moved by air or a liquid.

VORTEX—A whirling movement or mass of liquid or air.

W

WORM GEAR—A gear with teeth that resemble a thread on a bolt. It is meshed with a gear that has teeth similar to a helical tooth except that it is dished to allow more contact.

SUGGESTED READINGS

FUNDAMENTALS OF SERVICE: *Bearings and Seals;* John Deere Service Training, Dept. F.; John Deere Road; Moline, IL 61265

FUNDAMENTALS OF SERVICE: *Belts and Chains;* John Deere Service Training, Dept. F.; John Deere Road; Moline, IL 61265

FUNDAMENTALS OF SERVICE: *Electrical Systems — Compact Equipment;* John Deere Service Training, Dept. F.; John Deere Road; Moline, IL 61265

FUNDAMENTALS OF SERVICE: *Engines — Compact Equipment;* John Deere Service Training, Dept. F.; John Deere Road; Moline, IL 61265

FUNDAMENTALS OF SERVICE: *Fuels, Lubricants and Coolants;* John Deere Service Training, Dept. F.; John Deere Road; Moline, IL 61265

FUNDAMENTALS OF SERVICE: *Hydraulics — Compact Equipment;* John Deere Service Training, Dept. F.; John Deere Road; Moline, IL 61265

FUNDAMENTALS OF SERVICE: *Identification of Parts Failures;* John Deere Service Training, Dept. F.; John Deere Road; Moline, IL 61265

Small Tractor Service Manual, Volumes 1 & 2, 7th Edition; Intertec Publishing Corporation; P.O. Box 12901; Overland Park, KS 66212

INDEX